Re-Centering Women in Tourism

ANTHROPOLOGY OF TOURISM: HERITAGE, MOBILITY, AND SOCIETY

Series Editor

Michael A. Di Giovine (West Chester University of Pennsylvania)

Mission Statement

The Anthropology of Tourism: Heritage, Mobility, and Society series provides anthropologists and others in the social sciences and humanities with cutting-edge and engaging research on the culture(s) of tourism. This series embraces anthropology's holistic and comprehensive approach to scholarship, and is sensitive to the complex diversity of human expression. Books in this series particularly examine tourism's relationship with cultural heritage and mobility and its impact on society. Contributions are transdisciplinary in nature, and either look at a particular country, region, or population, or take a more global approach. Including monographs and edited collections, this series is a valuable resource to scholars and students alike who are interested in the various manifestations of tourism and its role as the world's largest and fastest-growing source of socio-cultural and economic activity.

Recent Titles in the Series

Re-Centering Women in Tourism: Anti-Colonial Feminist Studies, edited by Frances Julia Riemer

Transnational Yoga at Work: Spiritual Tourism and Its Blind Spots by Laurah E. Klepinger

Medical Tourism & Inequity in India: The Hyper-Commodification of Healthcare by Kristen Smith

A Sacred Vertigo: Pilgrimage and Tourism in Rocamadour, France by Deana L. Weibel

Women and Tourist Work in Jamaica: Seven Miles of Sandy Beach by A. Lynn Bolles

Encounters across Difference: Tourism and Overcoming Subalternity in India by Natalia Bloch

Study Abroad and the Quest for an Anti-Tourism Experience edited by John Bodinger de Uriarte and Michael A. Di Giovine

Re-Centering Women in Tourism

Anti-Colonial Feminist Studies

Edited by Frances Julia Riemer
Foreword by Florence E. Babb

LEXINGTON BOOKS
Lanham • Boulder • New York • London

Published by Lexington Books
An imprint of The Rowman & Littlefield Publishing Group, Inc.
4501 Forbes Boulevard, Suite 200, Lanham, Maryland 20706
www.rowman.com

86-90 Paul Street, London EC2A 4NE

British Library Cataloguing in Publication Information Available

Library of Congress Cataloging-in-Publication Data Available

ISBN 978-1-66690-106-1 (cloth)
ISBN 978-1-66690-107-8 (electronic)

Contents

List of Figures

Foreword

Setting the Stage

Florence E. Babb

I was teaching my undergraduate course on the anthropology of travel and tourism in spring 2020 when the COVID-19 pandemic descended on the world. The irony of studying the nuances of travel at a time when students were heading home and unable even to go to campus was not lost on me or my students. Nonetheless, we cheerfully turned what we thought would be a short-term situation into an opportunity for imagining travel in the future and questioning how we could bring more critical attention to the often taken for granted entitlement and self-actualization of tourists hailing from the global North. As I write this foreword nearly two years later, we are still living the pandemic, which has become the "new normal," at least for now. And as I prepare to spend a part of my sabbatical semester in Ecuador, I know I will pack up plenty of masks and be ready for social distancing, as much as that is possible when traveling.

In this brave new world of calculated risk-taking, Frances Julia Riemer is bringing out a welcome collection of essays that reframe travel and tourism from a critical feminist and anticolonial standpoint. The volume is a robust mix of contributions that in their different ways seek to go beyond earlier efforts to theorize and produce ethnographic accounts of women as hosts and guests in tourism sites around the globe. No longer content simply to stake claims for the difference that gender makes in the travel experience, these essays complicate narratives of travel by examining the deeply gendered and racialized ways in which both women and men often engage in the tourism encounter. The authors are acutely aware of the ways that travelers, particularly those from the global North who head for the South, unwittingly reproduce colonialist tropes of exploration, rescue, and redemption. Going beyond broad notions of the "tourist gaze," they consider how even an affective, caring, or "humanitarian" gaze can signal the reinscribing of relations of power between tourist and toured. They question how such performances of feminized qualities on the part of both guests and hosts can mobilize and animate touristic encounters, something I have observed in my own research on Latin American tourism over the years (Babb 2011, 2018).

In many ways, this volume's state-of-the-art assessment of work on "women in tourism" reflects currents in feminist anthropology more broadly over the last several decades. From initial efforts to add women to the narratives in our discipline to derive a more complete story of human development, to a turn toward recognizing gender as relational and socially produced, we are now seeing a more reflexive, critical feminist anthropology that engages gender in relation to other forms of difference in complex webs of power (Babb 2013). This should not be understood as an inevitable "progress tale" however, as we are constantly running up against our own limitations and lingering colonialist ways of viewing and being in the world. As such, contributors to this anthology offer much-needed, often self-critical, discussion of their interventions in tourism research (Karen Stocker and A. Lynn Bolles, on working with local collaborators in Costa Rica and Jamaica, respectively) and in travel itself (Siân Stephens, on her defiant entitlement and postcolonial guilt on the streets of Morocco).

This collection has the value of examining women in tourism both as local workers and as international visitors. In the first category, we read about those who provide the feminized and precarious labor of Spanish-language instructors in Guatemala (Sarah Becklake), and of the "monetized mothering" of Costa Rican women whose cooperatives make jam and apparel and who now offer homestays to tourists (Karen Stocker). We discover the innovative culinary tourism of Arab women minorities living in Israel who practice "culinary militarism" via home-style cooking to insist on their legitimate place in the Jewish-Israeli nation (Lindsey B. Pullum). We also learn of the efforts of women basket weavers of Botswana whose cultural heritage and commodified art is now a dying tradition (Frances Julia Riemer). All these women persevere against the odds when they are confronted by the forces of neoliberal globalization, though their future is uncertain.

On the visitors' side, too, we find that women's travel experiences run the gamut from voluntourism, to pursuing intimate relations and collecting "travel sex" stories, to "lingering" and becoming resident ex-pats living abroad. Tourist volunteers are largely though not exclusively white and female and often direct their activities to caring for poor people of color, seeking safe spaces for their own self-transformation, as in the case of a Canadian NGO sending travelers to Guatemala (Amy Kipp). Seemingly at the other end of the spectrum are those women traveling to India and elsewhere who delight in the liminal space of travel where they seek out sexual liaisons that make for thrilling narratives back home (Emily Falconer). And then there are the travelers who find it hard to leave places they deem to be a sort of paradise, where they can afford to live well on very little, as in the case of hard-partying ex-pats in Costa Rica (Kris Maksymowicz).

Taken together, these narratives reveal locals persevering and making the best of opportunities to gain a living through tourism development and of travelers whose objectives are frequently to gain new experience through their association with locals, taking away too-easy understandings of North-South differences. The authors' critical feminist and anticolonialist perspectives provide deeper insights into these commodified cultural exchanges and desires to know and even consume the "other." Through intimate encounters—whether economic, cultural, or corporeal—differences are often reinscribed and visitors return home feeling empowered, having consolidated their sense of earned privilege and entitlement. Gender, class, and racial difference are all salient in tourism settings, yet while women hosts and tourists alike may experience degrees of subordination to men, women travelers often capitalize on their class, racial, and national status to offset any efforts by local men to upend their privilege.

In her introduction to the chapters that follow, Riemer comments on problems stemming from the excess of tourism to certain destinations before the pandemic and the fierce reactions of some locals who have felt overrun by seasonal invaders. We may ask what is ahead for tourism, if and when restrictions on travel and fears of contracting new COVID variants diminish. Will there be a swift return to travel-as-usual with the same charges of tourism's excess and its harms to the environment and quality of life for many cities and communities? Will well-heeled travelers rush in to take advantage of places that were recently off-limits, bringing with them the attitudes and behaviors that have long offended local hosts? Or have we been chastened to consider the fragility of global ecologies and human-nonhuman relations, to seek new means of knowing the world—sometimes even forgoing opportunities to travel to distant places where we leave behind large carbon footprints? Will women travelers, even those with humanitarian impulses to "lend a hand," come to consider the high cost to local hosts of providing for tourists' care and safety while they seek self-fulfilling experiences abroad? Will they recognize the long-term implications of indulging in short-term intimacies, and their own racialized, colonialist desires? And just as importantly, will local hosts roll out the welcome mat once again or push back against such practices that reproduce old familiar hierarchies?

This volume, in re-centering and re-imagining women in tourism, will serve us in important ways as we contemplate possible futures of travel. The feminist scholars who have contributed to this project offer critical reflections on their own practice of carrying out research on tourism, their political engagement and social commitment to better futures for those who depend on tourism for a livelihood, and for those seeking alternative and more just forms of travel. A feminist practice of care for relations on the planet and for addressing geographies of difference expressed through travel is surely an excellent starting point. Ultimately, we are challenged to work ever more critically to decolonize our tourism research so that we develop pluralist conceptualizations, thinking with theorists from the South to frame inclusive and far-reaching understandings of tourism in a mobile world (Chambers and Buzinde 2015). This collection is a crucial provocation in setting us on that path.

REFERENCES

Babb, Florence E. 2018. *Women's Place in the Andes: Engaging Decolonial Feminist Anthropology*. Berkeley, CA: University of California Press.

———. 2013. "Feminist Anthropology." In *Theory in Social and Cultural Anthropology*. Edited by Jon McGee and Richard L. Warms, vol. 1, 258–262. Thousand Oaks, CA: Sage.

———. 2011. *The Tourism Encounter: Fashioning Latin American Nations and Histories*. Stanford, CA: Stanford University Press.

Chambers, Donna, and Christin Buzinde. 2015. "Tourism and Decolonisation: Locating Research and Self." *Annals of Tourism Research*, vol. 51: 1–16.

Acknowledgments

This book is a group project, and I thank Florence Babb, Siân Stephens, Amy Kipp, Emily Falconer, A. Lynn Bolles, Sarah Becklake, Kris Maksymowicz, Lindsey B. Pullum, and Karen Stocker for their contributions. I applaud their creativity, their feminist sensibilities, and trenchant analysis, and am indebted to them and their willingness to engage with this anticolonial feminist project of re-centering. We also express our deepest gratitude to the communities in which we work, and to the women who talk with us about triumphs and challenges of global tourism. I extend my thanks to Michael DiGiovine, Lexington series editor, who has been steadfast supporter of the work and a tireless advocate of the anthropology of tourism. I am also grateful to Ryan Dradznyski, assistant editor at Lexington Books, who checked our indentations and images and fashioned the chapters into a complete manuscript.

This book is also a personal project. *Re-Centering Women in Tourism* rose from my need to address ideological contradictions in my own international forays and professional experiences and later, my field work. I acknowledge all the *feminists*—Gayatri Spivak, Chandra Mohanty, Maria Lugones, Maria Mies, Rafia Zakaria, and so many others—whose writing helps us sort through the implications of our own travels and our unwitting participation in neo-colonial re/production. I am also indebted to my colleague, Dr. Sanjam Alhuwalia, as well as other *good trouble makers* on campus. I am thankful for our engagement—in conversation, text, and classroom—in thinking through the implications of anticolonial work in our writing, teaching, and practices.

My first grandchild, Max Dennis Rowan-Hattem, was born while this book was in press. I trace my love of travel to trips I took as a child with my own grandmother. Now a grandmother myself, I dedicate *Re-Centering Women in Tourism* to Max, and I look forward to traveling this glorious world together.

Frances Julia Riemer, PhD
Sedona, AZ

Introduction

Anticolonial Feminist Studies of Tourism

Frances Julia Riemer

A MOMENT OF PAUSE—DECEMBER 2022

We draft these chapters as the world has re-opened. COVID-19 shut down tourism and kept us home for more than 15 months. The closure was a radical change; before COVID-19, the tourism industry was "overbooked" (Becker 2013). Overtourism was the topic of 2018 at both a UN co-organized gathering of more than 60 tourism ministers and private-sector leaders and at ITB Berlin (Internationale Tourismus-Börse Berlin), the world's largest tourism convention (Pannett 2018). Host communities also pushed back. "Tourists go home" was ubiquitous graffiti in Barcelona and other tourist destinations around Spain (Burgen 2018). New Zealand's tourism boom, the result of the country's marketing of its "Lord of the Rings" Middle Earth scenery, became a cause of concern for its residents. The government imposed a "visitor tax" and deported at least one English family for their bad behavior (Graham-McLay 2019, A9).

But in 2020, the rapid spread of COVID-19 shut down tourism across the globe. Cruise ships were docked, planes were grounded, and restaurants and bars closed their doors. With movement limited, air quality improved, GHGs emissions decreased, water and noise pollution lessened, and the pressure on tourist destination was reduced (Owens 2021; Rume & Didar-Ul Islam 2020). Dolphins were reported to have wandered into Venice's Canals (Buckley 2021), and pumas were seen on the streets of Santiago Chile (McDonnell & Poblete 2020).

We write as the world has re-opened. COVID-19 vaccinations have become available, albeit at radically different paces around the world. Planes are again operating at full capacity, and cruise ships are scheduling departure dates. But even this opening is fragile, as one after another COVID variant rises. During this time of still guarded travel, we might remember that the historic global pandemic offers a moment of pause to rethink the ways we travel, the ways we congregate in place. A moment of pause to consider what is being called "the new normal" of travel.

Italy's recent ban on cruise ships and other "enormous" boats from Venice's canals and lagoons provides a stunning example of this reset. Spurred by the city's residents, environmentalists, and a UNESCO threat to put the city on the "World Heritage in Danger" blacklist, the shift signals "a move to protect a fragile ecosystem from the downsides of mass tourism" (Painigianni & Bubola 2021). Might the ban signal a paradigm shift in the ways we think about economic interests, environmental concerns, culture, and tourism?

This text is perfectly timed for this moment of reassessment. In *Re-Centering Women in Tourism: Anti-Colonial Feminist Studies*, we contribute to the conversation by offering a unique and long overdue theoretical frame on tourism. By positioning tourism as gendered neo-colonial practice and by centering women[1] and their/our experiences (e.g., as hosts, liaisons, vendors, performers, producers, and consumers) through lenses that encompass colonial histories and economics, we reframe the very presuppositions on which tourism initiatives are based. Our focus is on how racialized and gendered logics, informed by neo-colonial structures, policies, and practices, shape women's participation in tourism initiatives, practices, and policies. Addressing tourism as simultaneously empowering women and reproducing colonial hierarchies, we decolonialize the knowledge that typically grounds tourism policies and practices, and we help imagine sustainable and regenerative alternatives.

My ethnographic research with women basket weavers in Botswana's Okavango Delta over the past 25 years jumpstarted this edited project. Over more than two decades, profit from the sales of the women's baskets supported both Botswana's tourism handicraft industry and missionary activities in the region. But as expatriate church-based intermediaries and consumers from the Global North left the country for other postings, the market for the high-end baskets shrank to nothing and the women could no longer afford to weave (see chapter 7). Questioning how the basket weavers' experiences, both rewarding and challenging, reflect women's involvement in the tourism industry writ large, I invited both junior and senior scholars to join *Re-Centering Women in Tourism*, a project that interrogates tourism as gendered neo-colonial practice. The invitation resulted in the combined work of a group of gifted junior and senior feminist anthropologists who examine tourism initiatives that range in contexts from Guatemala and Costa Rica to

Israel and Botswana. Across chapters, we ask, what happens when we employ feminist anticolonial lenses in the study of tourism? What do tourism projects look like when we center and contextualize the lived experiences of women? With the shared goal of imagining sustainable and regenerative tourism possibilities, we pay particular attention to place (i.e., popular tourist locations), programs (i.e., language learning, volunteer tourism), practices (i.e., traveling alone, sexual interactions), and related educational and NGO initiatives that encourage and support gender equality and women's agency. Together we embrace the building of new knowledge and advocate for change.

Guided by a robust literature in the social sciences regarding research positionality that dates back to a reflexive turn in anthropology (Clifford and Marcus 1986) and a feminist mandate to locate self in research (Anderson 2020; Haraway 1988; Franks 2002), we also attend to our own situatedness in the scholarship and research sites. Given our relatively privileged backgrounds, we recognize that we far more closely resemble the wealth, status, and histories of tourists than of tourism's hosts and liaisons. We contemplate how our own identities, encoded historically by neo-colonial presuppositions, interests, and practices, shape the research. We examine biases specific to women researchers in tourism studies and reflect on our positionalities as researchers and scholars from the Global North (Porter and Schänzel 2018).

For myself and many of my colleagues, the work and the theoretical shift to anticolonial feminism is personal as well as epistemological. I am a former development worker with experience supporting women's projects in east and southern Africa, Southeast Asia, and Latin America, and was long committed to the idea of women's income generation in the tourism industry as empowerment strategy. But over time, I have become aware of post development theorists (Escobar 1994; Ferguson 1990; George 1997) who argued that the very idea of development, that is, prioritizing growth based on Western economic models, is most often harmful to both the land and its human stewards. As applied to women and tourism, the shift in orientation troubles the relationships—Global North and South, wealthy and poor, consumption and production—on which tourism initiatives are typically based. This shift in orientation is not easy; it compels questions about our engagement in and support of normative tourism-based activities. But it also forces us to consider other versions of tourism that are both sustainable and socially equitable.

THE SEARCH FOR A THEORETICAL FRAMEWORK: TRADITIONAL EXPLANATIONS

In 2019, *Women in Tourism* as a concept was all over the media. Tourist companies promoted their support of local women's businesses. South Africa's

August #IAmTourism campaign of photos, for example, highlighted the role of women in the country's tourism industry. Bloggers wrote about female focused travel companies, and cooperatives and universities began offering courses in women and tourism.

While the introduction into popular discourse may be relatively recent, *women in tourism* has been a focus of academic study for the past 50 years. At least since Sinclar's (1977) *Gender, Work and Tourism* and its comparative international perspective on tourism, work, and gender, scholars have investigated the ways that gender norms affect both the ways in which labor is supplied in the tourism sector and the nature of the demand for it. Subsequent research might be divided by at least three foci: querying gender as one of several variables in the sustainability and economic returns of tourism projects (Ferguson & Alarcón 2013, 2014), improving normative practices (Cole 2018; Lawless et al. 2020), and studying tourism projects designed specifically to promote women (Boonabaana 2014; Casellas & Holcomb 2001; Stonich et al. 1995).

The landmark *Global Report on Women in Tourism* 2010 was the first study of its kind to map the participation of women in the tourism sector worldwide. The report, focusing on employment, entrepreneurship, leadership, education and community, found that the majority of the tourism workforce worldwide is female—54 percent of people employed in tourism are women compared with 39 percent in the broader economy. The report also asserted that tourism offers women greater opportunities for leadership roles—23 percent of tourism ministers are female compared to 20.7 percent of ministers overall. In addition, both an investment in skills training for women and targeted interventions by public, private and civil society actors were found to promote decent work for women in tourism and lead to greater outcomes for gender equality.

By the 2019 World Tourism Organization's (WTO) *Global Report on Women in Tourism*, the industry was showing much promise for women. According to the report's findings, the majority of the tourism workforce worldwide is female, the gender wage-gap is smaller in the tourism industry than in other economic sectors, and tourism offers women greater opportunities for leadership roles. However, the report also noted that women are concentrated in lower status and lower paid jobs, often on a part time basis, in the industry.

The WTO report (2019) also highlighted a significant amount of sexual objectification of women working in the tourism industry, with sex tourism the most explicit example. In the broader industry, women were expected to dress in an "attractive" manner, to look beautiful (i.e., slim, young, pretty) and to "play along" with sexual harassment by customers. While tourism could play a role in disrupting these stereotypes, the industry is a reflection of gender roles in broader society (Möller 2012). And in practice, tourism's cash infusion has been found to reproduce, rather than disrupt gender, racial, and other cultural roles and hierarchies (Felton & Becker 2001; Lama, 1998; Munshi 2006; Pritchard 2014; Rinaldi & Salerno 2019; Vandegrift 2008; Wilkinson & Pratiwi, 1995). The tension of increased opportunities and reproduction of gender roles and salary disparities has been both constant and disconcerting in the research. But gender stereotypes, roles, and resultant inequities are not tourism-specific; in fact, the tourism sector appears to be yet another example of the gendering roles and gendered stereotypes.

Anticolonial Feminisms

We frame our research on women and tourism differently. Employing anticolonial feminist theories in our inquiry, we argue for "politics and epistemologies of location, positioning, and situating" that require us to locate tourism historically, economically, politically, and, most importantly, in the lives of women (Haraway 1988, 589). This project of epistemological decolonization of tourism is both vital and transformational. Drawing on Gayatri Chakravorty Spivak's (1988) focus on critical realignment of colonial power and knowledge, Maria Lugones's (2010) analysis of colonial hierarchies, Lila Abu Lughod's (2002) critiques of white saviors, and Chandra Talpade Mohanty's (1984) discursive constructions of third world women, anticolonial feminist scholarship, including both postcolonial and decolonial projects, opens and questions the presuppositions of dominant discourses employed to make sense of our worlds.

We use the term anticolonial feminism—a "theoretical and a political project that challenges imperialist and colonizing practices, past and present" (Mendoza 2016, 100)—to include both decolonial and postcolonial feminist analyses. Divided initially by colonial histories, postcolonial feminist scholars from the South Asian Subaltern Studies Group center their work in India, Nepal, and other former British colonies, while decolonial feminists from the Latin American Modernity/Coloniality Group originate and focus on Spanish and Portuguese colonization in the Americas. Particularities differ. Postcolonial feminists address epistemic violence and the subaltern (Mohanty 1991; Spivak [1985] 1988); decolonial feminists write of coloniality of power (Quijano 2010) and mestizo consciousness (Anzaldúa 1987). But across geographies, feminist scholars provided frameworks for analyzing power and knowledge as prescribed by coloniality, gender, and race.

Through this lens, tourism—both as "theoretical construct and as a perceptual mechanism"—is neo-colonial activity (D'Hauteserre 2004, 237). The industry mirrors "historical patterns of colonialism and economic dependency" (Burns 2004, 6–7), and global relations of domination and subordination are reflected in tourism's core/periphery relationships, representations of Other, and imaginaries and the colonial exotic (Kincaid 2000; Riemer 2020; Tucker 2019). Tourism is "an industry which, it could be argued, invites first world western tourists to consume third world places and people as pleasure products" (Aitchison 2001, 135). But tourism not only "invites"; the industry also normalizes this Global North/South relationship of consumer and consumed.

Our use of anticolonial feminisms not only reframes tourism as neo-colonial activity, but also decenters the white Western Eurocentric male experience of and perspectives on tourism. By centering women, we unabashedly embrace feminism as epistemic approach (Chambers, Munar et al. 2017). Feminist epistemologies unpack the role, nature, and effects of gender (in)equality in the industry. We write about the work of solidarity across role, status, and location, and ask what travel looks like that does not involve gender objectification. As Mohanty (2003, 2) reminds us, "feminism without borders" does not mean an erasure of differences or borders, but instead, an awareness that lines cut through and across "nations, races, classes, sexualities, religions, and abilities are real." That awareness, Mohanty writes, "must envision change and social justice work across these lines of demarcation and division."

THE CHAPTERS

The authors of the chapter-length ethnographies in this text draw on anticolonial feminisms to re-center women in tourism. In the re-centering, they offer unexpected, and at times, shocking ways of seeing tourist encounters. As readers, we recognize the participants and the settings of the ethnographies. Perhaps we've had a volunteer tourism experience, studied a language internationally, traveled alone, or bought a handmade craft from a tourist shop. But when authors use feminist anticolonial lenses to interrogate the ways we travel, the re-centering of those experiences is often disorienting. Perhaps then, this is as much a warning about, as it is an introduction to the chapters.

Re-Centering Women in Tourism is organized into three sections that represent ways of experiencing tourism. The first examines touristing—the going to other places for recreation and education. Siân Stephens provides a jumping point for a radical re-centering. In "Who Invited the women?" Stephens addresses solo female travel in contexts where solo female travel is deemed culturally inappropriate. She frames travel as colonial intrusion, and questions what equal rights to equal access to public space means to and for travelers. Next up, Amy Kipp takes on volunteer travel. With a focus on women's experiences in Guatemala, Kipp uses an intersectional feminist and an anti-colonial approach to conceptualize the volunteer tourist bubble.

Centering on the experiences and narratives of women volunteers, Kipp argues that rather than ensuring the authentic experience touted by volunteer tourism providers, the bubble offers security, safety and predictability in ways that reproduce racialized, gendered and colonial practices, beliefs, and policies. In "Skanky Stories," the final chapter in the touristing section, Emily Falconer broaches the raunchy and typically taboo sex stories of female backpackers. Framing "travel space" as unlimited freedom for women backpackers and their stories as liberation "from the confines of femininity without the emotional burden of embodied and sexual shame," Falconer illustrates that the women's stories simultaneously represent a rite of passage, sexual agency, and the "postcolonial privilege of backpacking" that is gendered, classed, and racialized. By re-centering on women's experiences and narratives, all three chapter authors remind us that travel is not neutral. Focusing on women traveling—alone, as volunteers, as backpackers–Stephens, Kipp, and Falconer trace the ways gender, race, and class intersect to mirror and reproduce colonial tropes and hierarchies in unexpected ways.

The second section of *Re-Centering Women in Tourism* focuses on hosting—the schools, communities, and individuals who manage and most often, commodify tourism. Here again is the disorientation that comes with our re-centering. A. Lynn Bolles focuses on women's work in Negril, Jamaica's "capital of casual," where gender, skin color, class, and access to education combine in the country's tourism marketplace. She introduces us to women tourist workers across wage earning categories, paying particular attention to higglers, and their business model, employed by poor and working-class women in Negril's tourism. Sarah Becklake's focus is on the experiences of female Spanish-language teachers in a school for visiting students in Antigua, Guatemala. Pedagogical tourism, or tourism undertaken in the pursuit of "educational" objectives, reflects a growing sector of the tourism industry and is of particular interest to younger Millennial and Gen Z travelers (WYSE Travel, 2018). But Becklake describes the unfolding of intersecting imperial, race, gender, and class relations and the respective processes of exclusion, domination, exploitation, and resistance as Spanish school directors/owners compete to attract and satisfy international—primarily white Western—language students. Then, in compelling prose, Kris Maksymowicz invites us to a Costa Rican beach town, where local resistance to expatriates takes the form of unexpected verbal outbursts, back talk that is quickly denied through unforgetting. Maksymowicz examines these "minor gesture[s]" as surfacing from/through the confluence of colonialism, racism, and the neoliberal tourism-industrial complex in Costa Rica that positions locals as antithetical to the country's Pura Vida brand and vibe. Frances Riemer (this author) closes the section with an analysis of tourism and women's handicrafts. Many of us have purchased a handmade souvenir to take a bit of our travels home with us. Riemer illustrates that this commoditization of women's work for a tourist market is all too often founded on a white liberal feminist model of economic empowerment, reflecting colonial relationships, roles, and representations. With a focus on female language teachers, restaurant cleaners, and handicraft producers, authors highlight the women's rationality, agency, and at times, resistance, in the face of insurmountable neo-colonial structures, policies, and practices.

The final section of *Re-Centering Women in Tourism* provides examples of equitable, sustainable tourism projects owned and managed by women. We begin with Lindsey Pullum's "My Mother's Recipe, My Nation's Narrative," and an introduction to Maisa, who owns and operates the only woman-owned Druze restaurant in Israel. Pullum's focus is on what she calls "culinary militarism," the ways Druze women working in culinary tourism engage with the Israeli state's colonial rule and their own cultural identities. Pullum reminds us of the power of women's knowledge and foodways in performing and experiencing "belonging" both in Israel's tourism economy and as a member of the country's minority community. We follow with Karen Stocker's participatory research with women's cooperatives in Costa Rica. In "Entrepreneurial Domesticity," Stocker tells the story of a more than two-decades long feminist participatory action research (FPAR) project in which she participated, first as an undergrad researcher and then as professor, bringing her own undergraduate students to the rainforest ecotourism destination. Stocker traces the FPAR research project as it documents and supports the evolution and growth of the women's tourist-oriented businesses, and as the women began to lead the local tourism industry "deliberately and conscientiously." The alternative projects described in this section disrupt tourism's neo-colonial relationships and support local ownership of and participation in the tourist economy. To return again to Mohanty, they help us envision, "change and social justice work across these lines of demarcation and division" in our own travel practices and in the tourism industry writ large.

REFERENCES

Abu-Laghod, L. (2013). *Do Muslim women need saving?* Harvard University Press.

Aitchison, C. (2001). Theorizing other discourses of tourism, gender, and culture: Can the subaltern speak (in tourism)? *Tourist Studies* 1:2, 133–147.

Anderson, E. (2021). Feminist epistemology and philosophy of science. In Edward N. Zalta (ed.), *The Stanford Encyclopedia of Philosophy.* https://plato.stanford.edu/entries/feminism-epistemology/.

Buckley, J. (2021, March 23). These dolphins took a day trip up Venice's Grand Canal. *CNN Travel.* Retrieved from https://www.cnn.com/travel/article/venice-canal-dolphins/index.html.

Chambers, D., Munar, A.M., Khoo-Lattimore, C., & Avital, B. (2017). Interrogating gender and the tourism academy through epistemological lens. *Anatolia* 28:4, 501–513.

Clifford, J., & Marcus, G.E. (1986). *Writing culture: The poetics and politics of ethnography.* University of California Press.

Cole, S. (2018). *Gender, equality, and tourism.* CABI.

D'Hauteserre, A.M. (2004). Postcolonialism, colonialism, and tourism. In Lew A.A., Hall, C.M., Williams, A.M. (eds.), *A Companion to Tourism*, 235–245. Blackwell Publishing.

Ferguson, L., & Alarcón, D.M. (2014). Gender and sustainable tourism: Reflections on theory and practice. *Journal of Sustainable Tourism* 23(3), 401–416.

Ferguson, L., & Alarcón, D. M. (2013). Integrating gender into sustainable tourism projects. *Equality in tourism: Creating change for women.* Equality in Tourism.

Franks, M. (2002). Feminisms and cross-ideological feminist social research: Standpoint, situatedness and positionality—Developing cross-ideological feminist research. *Journal of International Women's Studies,* 3(2), 38–50.

Haraway, D. (1988). Situated knowledges: The science question in feminism and the privilege of partial perspective. *Feminist Studies*, 14(3), 575–599.

Kincaid, J. (2000). *A small place.* Farrar, Straus, and Giroux.

Lawless, S., Cohen, P., McDougall, C., Orirana, G., Siota, F., & Doyle, K. (2019). Gender norms and relations: Implications for coastal livelihoods. *Maritime Studies,* 18, 347–358.

Mcdonnell, P.J., & Poblete, J. (2020). Mountain lions on the streets: Chilean capital sees influx of big cats amid lockdown. *Los Angeles Times*, September 5. https://www.latimes.com/world-nation/story/2020-09-05/mountain-lions-on-the-streets-chilean-capital-sees-big-cat-influx-amid-lockdown.

Mendoza, B. (2016). Coloniality of gender and power: From postcoloniality to decoloniality. In Disch, L.J. & Hawkesworth, M.E. (eds.), *The Oxford Handbook of Feminist Theory* 100–121. Oxford University Press.

Mitra, D. (2020). *Indian sex life: Sexuality and colonial origins of modern social thought.* Princeton University Press.

Mohanty, C.T. (2003). *Feminism without borders: Decolonizing theory, practicing solidarity.* Duke University Press.

Owens, B. (2021). The positive and negative impacts of COVID on nature. *Hakai Magazine,* May 27. https://www.smithsonianmag.com/science-nature/positive-and-negative-impacts-covid-nature-180977840/.

Pianigiani, G., & Bubloa, E. (2021). Italy's Government to ban cruise ships. *New York Times*, July 13. https://www.nytimes.com/2021/07/13/world/europe/venice-italy-cruise-ship-ban.html.

Porter, B.A., & Schänzel, H.A. (2018). *Femininities in the field: Tourism and transdisciplinary research.* Channel View Publications.

Quiano, A. (2010). Coloniality and Modernity/Rationality. *Cultural Studies,* 21:2, 168–178.

Rume, T., & Didar-Ul Islam, S.M. (2020). Environmental effects of COVID-19 pandemic and potential strategies of sustainability. *Heliyon,* 9. https://www.ncbi.nlm.nih.gov/pmc/articles/PMC7498239/#:~:text=The%20global%20disruption%20caused%20by,parts%20of%20the%20world.

Spivak, G. (1988). Can the subaltern speak? In Nelson, C. & Grossberg, L. (eds.), *Marxism and the Interpretation of Culture.* Macmillan.

Tucker, H. (2019). Colonialism and its tourism legacies. In D.J. Timothy (ed.), *Handbook of Globalisation and Tourism*. Edward Elgar Publishing.

WYSE Travel. (2018). *New Horizons IV: A global study of the youth and student traveler*. WYSE Travel Federation.

NOTE

1. Recognizing feminist histories of essentializing the term "woman" and the ways that gender has been interrogated, authors also understand "woman" as defined and conceptualized differently across cultures and to include not only "cis" women, but trans women, femme/feminine-identifying genderqueer and non-binary individuals as well.

SECTION I

Touristing

Chapter 1

Who Invited the Women?

The Double Bind of a Culturally Respectful Female (or Feminist?) Traveler

Siân Stephens

INTRODUCTION

The phenomena of travel and tourism is riddled with boundaries, and there are often quite messy implications when these boundaries are crossed or blurred (Aitchison 2001). In this chapter I consider the challenges presented by the endeavor of anti-colonial feminist travel in a post-colonial context, by examining the risk of neo-colonialism posed by solo female travel in contexts where solo female travel may be deemed culturally inappropriate. Here I use the term "imperial" to refer to "expanding a society's interests abroad" (Nash 1989), "neo-colonialism" as the pursuance of imperial interests via economic and social means (Huggan 1997; Wejesingh et al. 2019) and "cultural imperialism" as the homogenization of culture resulting from neo-colonial influences (Sarmela 1977). The term "anti-colonial feminism" refers to "a theoretical and a political project that challenges imperialist and colonizing practices, past and present" (Mendoza 2016, 100). It is often applied in work relating to the pursuance of a feminism which resists both gender-based and colonial oppression (e.g., Codorova 1998, Sunseri 2000), but in this chapter I am applying the term to a different but related phenomena. I consider not the pursuance of feminism in a way which resists the imposition of colonialism on the subject, but the pursuance of feminism in a way that resists the imposition of colonialism by the subject; in this instance, by the white, Western woman traveler. The imperialistic nature of tourism, exemplified by the expectation that the host make any adjustments required to make the guest feel comfortable, has been well-explored (e.g., Aitchison 2001; Nash 1989). Once our consciousness of these dynamics has been raised, it is difficult to pursue tourism without efforts to mitigate our imperialistic influence on the destination. This we may do by perusing authentic or legitimate travel (Mehmetoglu 2004), minimizing the demands we make of our host by offering to participate in their existence, rather than requiring them to adjust to our presence. However, such an endeavor in fact risks further intrusion, commodifying the host culture itself and drastically increasing the intrusion of the visitor on the destination (Nash 1989). There is much excellent work addressing implications of authentic travel (Brown 2013; Yeoman et al. 2007; Kolar & Zabkar 2010; Paulauskaite et al. 2017), but the implications of this practice when undertaken by women—particularly "unaccompanied" (meaning unaccompanied by a man) women, has not received adequate attention, despite the acknowledgement that women who travel face and pose unique challenges (Wilson & Little 2005). As I argue in this chapter, many of these challenges result from differences in attitudes to gender between the host culture and that of the visitor. In contexts where there are cultural restrictions in place preventing or discouraging women's pursuits of leisure in public there are neo-colonial implications where solo or unaccompanied women visitors seek leisure tourism.

Those of us who are committed to reducing the damage done to and promoting the rights of others are likely to hold values relating to cultural, ethnic, racial, and gender equality; we are likely to believe that women have the same entitlement to travel as men, and that cultures which differ from our own are entitled to our respect. Solo female travel is often presented as a way of exercising women's rights to equality; as an empowering, feminist act where one can prove one's capabilities and competence by overcoming the obstacles which travel presents. Contemporary discussions in this vein take place in blogs, online travel advice, and discussion fora. The advice offered in mainstream guidebooks (including the online versions) offers a valuable insight into the mainstream discourse regarding female travel (Jeffrey & Stephens 2021), and the existence of special sections in mainstream travel guides tells us a) that the advice in the book would otherwise be insufficient for women travelers and b) that women are likely to face specific challenges when traveling. This advice often focuses on how to respond to sexual harassment, as well as how to navigate countries where women's access to public space is limited. From this advice we can observe two tensions specific to the experience of unaccompanied women pursuing leisure tourism. The first tension is between the entitlement felt by the female traveler to exist in public spaces, even while visiting a place where this is not culturally appropriate. The second tension is between the Western conception of the solo female traveler as emancipated, independent, and sexually autonomous and the way in which she may be perceived (by the hosts) as, among other things, sexually available, often times resulting in attention which acts to limit the freedom of the traveler. Both tensions result from a friction between the entitlement of the traveler and the restrictions of the destination, and both tensions are gendered in that they relate to challenges facing women travelers because they are women.

A wide range of tourists and travelers face discrimination in the form of challenged access to public space, and it is not suggested here that white women are the primary victims of this form of exclusion. Valuable research has been done on the presence of racism in tourism (e.g., Benjamin & Dillette 2021; Lee & Scott 2017), the challenges facing the LGBTQ+ community (e.g., Ong et al. 2020), travelers with disabilities (e.g., Agovino et al. 2017) as well as obese travelers (e.g., Small & Harris 2012). However, in this chapter I focus specifically on the challenges which are pointed at women because they are women.

My academic interest in gender, travel, and tourism arose in response to my personal experiences as a woman who travels. In particular, I started to write about experiences which I found challenging, specifically those experiences which highlighted an internal conflict between tolerance and entitlement (Stephens 2020). My experiences are those of a white British woman, and my reflections and experiences speak to this particular set of baggage, but incompatibilities such as these are likely to be present for any woman who travels, although the conceptions themselves will differ. In this chapter I consider possibility of women perusing leisure in public as a colonial intrusion, and the implications of attitudes to sexual harassment when traveling abroad. These two issues draw our attention to the particular challenges facing women who want to travel independently to culturally distant locations, while showing respect to her hosts. These specific challenges are gendered, but they speak to the wider issue of the pervasive colonial legacy present of some forms of travel and tourism.

In order to address the challenge posed by anti-colonial women's travel I refer to advice offered to women travelers in mainstream travel guidebooks, specifically the Rough Guide and the Lonely Planet. Guidebooks shape not just what is worthy of seeing in a tourist destination but also how tourists should behave when they are there (Dann 2003). These texts offer insight into the dominant travel narratives of Europe and North American cultures, and provide useful data with which to explore attitudes to travel and to gender. As documents produced for public purposes, the text of travel guidebooks lends itself to discourse analysis, (Wodak 2011), which allows us to "identify and understand institutionalized patterns of knowledge that govern the formation of subjectivity" (Arribas-Ayllon & Walkerdine 2019,

110). Using a Foucauldian approach, I have reviewed this advice in order to better understand contingent rules governing the behavior of women travelers and socially permitted attitudes toward the hosts and destinations. It should be noted that this approach allows a critical analysis of socially embedded discourse and of the societal norms which can be observed and understood through text intended for public consumption; it is not a critique of the writing or of the writers.

PUBLIC PRESENCE AS COLONIAL INTRUSION?

Attitudes concerning appropriate behavior in public space vary from society to society (Cresswell 1996), and this is particularly relevant in relation to attitudes to gender and access to public space. For example, the Rough Guide advice to women travelers in Morocco advises "the street is strictly the man's domain" (Rough Guide 2021), raising questions about how the woman traveler is to get about if using the streets is culturally inappropriate. Within Western societies, women's use of public space is not unproblematic, and much has been written about the different ways in which our use of space is discouraged, from the use and threat of violence (Valentine 1989) to the way in which public space is designed (Boys 1985). However, it is a characteristic of modern Western feminism that women believe themselves to have equal rights to equal access to public space (despite the fact that this access is still hindered to a greater or lesser extent). Indeed, the growth in women's solo and unaccompanied travel can be seen as a consequence of this belief and an example of our efforts to stake claim to space which has historically been understood as belonging to men (Elsrud 2001; Falconer 2011). In this way, women's travel can be seen as transgressive, as an act of resistance against traditional gender roles and historical limitations to our mobility. These kinds of transgressive acts can be admirable and valuable; transgression has the effect of illuminating the lines of what is considered proper (Cresswell 1996) and therefore play an important role in highlighting inequality, injustice and discrimination.

However, in the case of solo female leisure travel it is not clear exactly which boundary is being transgressed. One possibility is that I, as a white British woman traveling alone, am transgressing the historical attitudes to gender and mobility of my own culture, which have historically dictated that travel is for men or accompanied women, but not for women alone. This is relatively unproblematic; the power dynamic between me and my culture and between women and men in my culture is one where I pose little risk of domination and therefore my transgression is at most, noble and at least, harmless. However, to view my transgression in these terms is to view travel through a colonial lens—in which all destinations "belong" to the men of my culture. Through this lens my act of independent travel transgresses the sexism of my own culture, within space belonging if not to me, then to my people. Adopting an anti-colonialist perspective however, would reflect a reality where the destination "belongs" to the host, and through this lens my transgression is against the host culture, not against my own. Such a transgression is less readily defensible in the many destinations where the power dynamic is more complicated. As a woman visiting India I am at risk from discrimination and domination due to my gender, but also at risk of discriminating and dominating as a result of my race, ethnicity, and status as tourist. My transgression into public space is not necessarily (or only) an important act of resistance, but may in fact be an act of cultural imposition.

This is a perspective I explored in an autoethnography on two trips I took to Morocco, where I encountered what I felt to be a high level of sexual harassment and resistance to my presence on the streets of the medina in Marrakech. In this work, I reflected on the advice offered in the Rough Guide that the "the street is strictly the man's domain" (Rough Guide 2021) and wondered at my willingness to violate a clear cultural norm of the host country as well as my assumption that this violation would be tolerated. This expectation was, at least in part, borne from a sense of superiority and entitlement—I believed that my cultural value of gender equality in the form of access to the public sphere was superior to the Moroccan culture where streets are men's domain. I also believed that even if this superiority was not accepted by the hosts, my violation of the cultural norm should be tolerated—welcomed even—because of my tourist status, and the racial and financial dominance implied by this status. This, I argue, is a relatively common attitude, and one which is reflected in much of the travel literature, including travel guides and blogs. Travel guides generally advise that the local culture be respected, but stop well short of advising against travel where one's presence would be culturally inappropriate. The strongest discouragement I have encountered is in the Lonely Planet advice to female travelers in Saudi Arabia, which states:

> Saudi Arabia is rightly considered one of the most difficult countries in which to travel if you're a woman by yourself. The strict segregation of the sexes and the need for a male chaperone leads to obvious limitations on freedom of movement . . . Furthermore, it would be impossible to find places to eat or sleep, as most restaurants outside of big towns are men-only affairs, and hotels in these areas will be reluctant to rent a room to a woman by herself . . . While Saudi Arabia is becoming far more comfortable for Saudi women, there is still a long way to go before the same can be said for female travellers to the country. (Lonely Planet 2021i)

While this advice is likely to discourage many solo female travelers, there is nothing in the advice which suggests an unaccompanied women should not go to Saudi Arabia because she is not wanted there. The tone in guidebooks such as this is consistently proactive and empowering; where obstacles such as the inability to eat in restaurants, strict cultural expectations regarding dress etc. are discussed, they are presented alongside advice as to how these obstacles can be addressed or overcome. The travel advice industry is based on an attitude of "if it is there we should go," and the advice targeted specifically at women is designed to ensure that they too can go wherever they want. The barriers described in the excerpt above relate to social constructions—they are barriers that the Saudi society has erected to make being an unaccompanied woman in public difficult, because the dominant culture in the country does not want unaccompanied women to be in public. And yet, the section goes on to advise women as to how these barriers can be navigated (e.g., to seek out "family only" hours at public attractions), in order that female travelers may find a way to not just visit Saudi Arabia, but to experience public life in Saudi Arabia—despite the obvious cultural incongruity of such experiences. Research suggests that in reality, visiting women are often effectively discouraged from accessing public spaces by barriers such as these (Brown & Osman 2017), but the overarching message in the guidebook advice is one which encourages solo female travel in destinations where traveling as a lone woman is culturally inappropriate. There is no sense that there are destinations where solo female travel should not be pursued.

Instead, it seems that there is a spectrum of culturally regulated behaviors; at one end are the behaviors we generally agree should be respected and at the other end are those we have implicitly or explicitly decided should be ignored or even resisted. Most guidebooks would advise travelers to cover-up in countries where the culture requires modest dress, many would advise that women cover-up a bit more than men. Beyond this, there is a range of culturally specific behaviors on which guidebooks offer useful insight which are relatively uncontroversial (e.g., which hand to use when eating, whether or not to shake hands or hug when meeting someone). This advice is generally offered in the spirit of making the visitor more comfortable, but it is also implied that such accommodations are important because they are polite and respectful. These behaviors are all at the "to be respected" end of the spectrum. At the "to be ignored or resisted" end of the spectrum lies behavior relating to women in the public space, women's mobility and women's leisure. It is beyond the scope of this chapter to explore how and where the line between "cultural behaviors to be respected" and "cultural behaviors to be ignored" is drawn. However, as a woman who has travelled to a number of countries where my presence can reasonably be assumed to have been disruptive, I can say that I went because I wanted to go and because I felt entitled to go. My conviction that gender discrimination was wrong and my belief that I should be able to go somewhere because I wanted to go there superseded any belief in the importance of respecting other cultures.

COLONIALISM, SEX, AND GENDER (OR SEXUAL HARASSMENT AS AN AUTHENTIC EXPERIENCE)

Sexual harassment is a particular form of discipline meted out to women who transgress the boundaries between the public and private. It is present in all countries and all cultures but has a particular affect in limiting women's mobilities within the context of independent travel as women navigate unfamiliar spaces (Wilson & Little 2008) and seek excitement, possibly in the form of risk (Falconer 2011). Sexual harassment has long been a feature of women's experiences in tourism (McElroy et al. 2008; Pritchard 2014) and attitudes to sexual autonomy and sexual availability are arguably the crucible of the cultural and colonial clash presented by solo and unaccompanied women's travel.

The inevitability of sexual harassment in some locations is addressed by mainstream travel guidebooks that advise women that the best response is to ignore it and, ideally, not to mind it (Jeffrey & Stephens 2021). It seems that sexual harassment is toward the "to be respected" end of the spectrum of culturally contingent behaviors. The causes and nature of the harassment will differ from place to place, and indeed from person to person, but again we can turn to the travel guidebooks to understand at least what are considered to be acceptable explanations of why such unwanted attention may arise. The dominant explanation offered is to do with a misunderstanding by the men of the host country of the nature of the visiting women. For example:

> In part this (harassment) is to do with Moroccan men's misunderstanding of Western culture and sexual attitudes, and the fact that some think they can get away with taking liberties with tourists that no Moroccan woman would tolerate. (Rough Guide 2021)

The suggested sources of this misunderstanding vary. Advice is offered regarding how women should dress, sometimes couched in terms of a strategy to avoid or minimize harassment (e.g., in the Lonely Planet advice for Cuba 2021c), but also offered as a way of seeming more approachable to local people and increasing the visiting woman's access to more authentic cultural experiences (as in the Lonely Planet Egypt 2021d). The advice to dress in a culturally appropriate way therefore seems to be offered in order to protect the host from the visitor, at least as much as the other way round.

Another common theme in advice to women sections refers to the influence of Western media, which misinforms local men about the likely sexual attitudes of visiting women. Hottola (2013) describes research conducted with women travelers from Europe in India, and considers the reasons for the sexual harassment they encounter. It is argued that the unwanted sexual advances are not necessarily a sign of disapproval, but result essentially from a misunderstanding, borne from the influence of films and the media. The sexual advances made in this context seem to arise from a genuine belief that consent has either been given or is at least very likely to be given, due in part to the signals being given by what the tourist is wearing and her behavior, filtered through the man's experience of "American soft-porn." The role of Hollywood in allowing men in sexually conservative countries to believe that Western women are likely to be sexually available is asserted in a number of guidebooks. The Lonely Planet advice for female traveler in Jordan reminds the traveler that "Stereotypes of foreign women based on Western films and TV convince some that all foreign women are promiscuous and will jump into bed at the drop of a hat" (Lonely Planet 2021f), and similar advice is offered to women visiting Iran. Here we have a complicated layering of assumptions and attitudes. A man's sense that he is entitled to sex or at least some form of sexual encounter is complemented by a belief that women who look like women from Hollywood films are sexually available to them, and compounded by the suggestion in Western writing that local men are sexually ignorant, passive receivers of media messaging. In this way cultural imperialism precedes the woman traveler, who finds herself plugged into an Othering narrative within which she is the Other (as well as the Otherer). The explanation offered by the guidebooks, that these assumptions are the result of cultural hegemony, put the woman tourist in a difficult position. If it is accepted that the men who are harassing her mean no harm but are simply confused by cultural imperialism, then the woman from a globally dominant culture would be wrong to get angry. The men who are making her feel uncomfortable are following the signals they believe she is sending, learned from her own culture's hegemonic media.

Other explanations as to why men in the host country may be confused about visiting women's sexual availability include a general misunderstanding of what it is for a woman to be in public alone. For example, there are suggestions that the confusion may arise from local men's beliefs that unaccompanied women are sex workers. In Lonely Planet Russia women are advised: "it's not uncommon for a woman dining or drinking alone to be mistaken for a prostitute" (Lonely Planet 2021h). It is unlikely that this belief arises from Hollywood films and more likely that it arises from local cultural practices, perhaps it is the case that in some countries unaccompanied women are often working women, and therefore an unaccompanied woman pursuing leisure is not recognizable. This is not a misapprehension based on cultural Othering as local men are simply applying the same rules they apply to local women to all woman. The cultural clash lies in the reaction of the female traveler, who may be tempted to object to being mistaken for a sex worker—certainly one might be a least irritated by having one's dinner interrupted with a business proposition.

Advice for visiting other countries suggests that simply being, or seeming to be unmarried is a signal which misleads men. For example, Lonely Planet's advice to women traveling to Bulgaria includes: "If you do attract unwanted advances, saying Az sâm omâzhena ('I am married') gives a pretty firm message" (Lonely Planet 2021b), and women visiting Jordan are told "Wear a wedding ring—this will add to your respectability in Arab eyes. A photo of husband and kids will clinch it" (Lonely Planet 2021f).

Mistreatment resulting from the belief that unmarried women are sexually available poses a particular challenge to the independent feminist traveler. In order to minimize the disruption of her presence and to better communicate her sexual unavailability, she is advised to pretend to be married. She is encouraged to suggest that she is sexually unavailable not because she is sexually autonomous, but because she has already been claimed by another man. This practice is likely to feel inauthentic for most women who would identify as feminists, and yet in some situations it is also perhaps the most culturally appropriate behavior. In these circumstances women are asked to subvert their own instincts to assert their autonomy in order to show deference to the host country's culture.

Further evidence of the conflict between autonomy and deference can be found in recommendations made to women travelers facing harassment. They are urged to ignore unwanted attention and to instead stay calm and respectful in the face of anything short of physical violence. It has been argued that this insistence on remaining beatific in the face of difficulties and discomfort is a gendered expectation, not necessarily of the host culture but of the travel and tourism culture (Falconer 2016). The "gift" of empowerment and independence comes at the price of subverted anger and tolerance of behavior we might find intolerable at home. Travel advice often reminds women that they should not find traveling particularly difficult, (Jeffrey & Stephens 2021) and women readers are reminded that, despite the litany of challenges presented in the "advice to women" section, other women have managed these challenges just fine. For example, the opening paragraph for the Lonely Planet's entry for Jordan begins: "Plenty of women each year have travelled through Jordan alone (this writer included) and have enjoyed the experience thoroughly," followed immediately by "That said, there are bound to be times when you will have male company in Jordan that you would rather do without. This may involve nothing more than irritating banter, proposals of marriage and declarations of love. On rarer occasions it will involve leering and physical contact—a grope in a bus, for example" (Lonely Planet 2021f). The implication is that sexual harassment is simply one of the obstacles to be overcome in the pursuit of authentic and adventurous travel, and information about such treatment is often tempered with a reminder of the importance of respect in the local culture. Advice for women travelers in Morocco, includes the assurance that "The key concept is 'respect,' something that most Moroccans hold dear, while also acknowledging the likelihood of sexual harassment" (Lonely Planet 2021g). Advice for travelers in Jordan is "to minimise harassment and to be respectful of local customs, it's imperative to dress appropriately, especially in small towns and rural areas." Entries for Argentina and Cuba reassure women that although they may receive unwanted attention, "Chivalry' is an important part of the local culture" (Lonely Planet 2021a, 2021c). The implication here is that the gender attitudes which the woman traveler might be finding difficult to navigate are in fact an authentic aspect of the local color; to mind too much would be to dismiss and disrespect the rich cultural experience of which she is in pursuit.

It must also be noted that in the tourists' home country, for example the UK or the US, it is unlikely that one would be encouraged to ignore sexual harassment because men are simply confused. Sexual harassment is a crime in many countries around the world, and it is generally accepted, outside the context of tourism, that being confused about a woman's sexual availability and/ or interest is not an acceptable excuse for harassment. The generosity afforded to perpetrators who pursue women travelers with unwanted attention seems to rest on assumptions of men's ignorance based on racial, ethnic, and cultural stereotypes. It may also be an attempt to persuade visiting women to minimize the disruption caused by their presence in an effort to reconcile what may be irreconcilable aims of anti-colonial feminist travel. As discussed in the previous section, the adventurous travel and travel advice industries are based on the belief that that if we as tourists want to go somewhere, we should go. The obstacles which the destination throws up can be seen as adventures to be had, not as deliberate constructions designed to discourage the presence of some visitors. It is also the case that travel advice is very often concerned with how to be a culturally respectful traveler, by informing the visitor of local norms and customs which may be adhered to with minimal inconvenience to the traveler. Given these aims and assumption, those giving advice to female travelers are in a difficult position. Allowing women to be angry or even to cause an active disruption in the face of harassment and discrimination would be to support culturally inappropriate behavior. However, to insist on tolerance and generosity is to demand a significant concession from the traveler, and one which is only (in this respect) asked of women travelers. The result of this conflict is that women are, on the one hand invited to take advantage of all the destination has to offer and, on the other hand encouraged to self-discipline their dress, behavior and emotional responses.

ANTI-COLONIAL FEMINIST TRAVEL?

My interest in this area arose from an internal conflict I experienced when traveling in some countries as an unaccompanied woman. I felt particularly uncomfortable in situations where I was in pursuit of an authentic cultural experience; when my intrusion on the host country was at its greatest. Staying in resorts and hotels is relatively easy, the cultural distance is dramatically minimized in these enclaves which are designed to cater to Western tourists who want to experience some but not all aspects of their destination. Travelers—adventurers who are seeking the opportunity to test themselves and overcome obstacles—do not stay in places where cultural distance is dramatically minimized. For open minded adventurers, cultural distance is the obstacle to be overcome, and as such, we seek out experiences where our presence is incongruous. And yet, incongruity, when sought in ignorance of local customs and culture, can easily lead to transgression. Examples of this include the British, Canadian, and Dutch tourists who posed naked on a sacred mountain in Malaysia (BBC News 2015) and the American tourists who took partially nude photos of themselves at the Ankor complex in Cambodia (Wong 2017). In these instances the transgressions were met with local outrage and international disapproval. There is some consensus that as far as possible local customs should be respected. Travel guidebooks, for example, would not recommend that non-Muslim tourists visit mosques where non-Muslim visitors are not welcome. Travel guidebooks do however routinely advise women on how to access public space in countries where unaccompanied women pursuing leisure are not welcome. These texts reveal a discourse absent any sense that one should not go where one is not wanted. An entire industry has sprung up to advise tourists on how to navigate "difficult" destinations and overcome the challenges which host cultures throw up to protect themselves from intrusion and imperialism. The invitation extended to women travelers to explore complex destinations is made by a dominant group of the dominant culture, not by the host culture.

The two cultural clashes described in this chapter highlight the particular challenges facing women travelers. The issues of access to public space and attitudes to sexual harassment draw our attention to the fact that there are, in public travel discourse, some cultural behaviors and practices which we are urged to respect and others which we are urged to circumvent. Sexual harassment is presented as a par-for-the-course element of women's travel. Women tourists are encouraged to dress appropriately and moderate their responses in an attempt to minimize the disruption that their provocative presence may elicit. Access to public space and the public pursuance of leisure however, are actively encouraged, and advice is offered as to how women can navigate the imposed restrictions. To some extent these attitudes reflect and reproduce the attitudes of the visitors' home culture (where their home culture is similar to that which produced the guidebook advice), adding evidence to the assertion that gender dynamics within tourism are manifestations of the gender dynamics of the societies which produce tourists (Pritchard 2014). The discourse considered in this chapter is produced by European publishers for Western audiences, and as such presents an insight into the gender attitudes of the West, and not of the host destinations referred to here. The compromise offered to women, of access to public space in exchange for unwanted attention and intrusion, is one which many European and North American women will recognize from home. This compromise, entailing as it does access to a space which women have previously been denied, invites women travelers to consider travel not only a right, but an important right, the exercising of which constitutes a political act asserting an entitlement to equality.

However, it is not for the Europeans to invite me to visit Jordan, India, Morocco and it is not for the West to offer the terms on which I may visit these places. The guest list is—or should be—at the discretion of the host, and an intrusive guest can cause unwelcome disruption. As a white British woman traveling in a post-colonial context, I naturally assume a position of colonizer. But I am also a woman, and in all cultures this puts me in a subordinate position, to a greater or lesser extent. While I may feel relatively comfortable asserting my position and my right to equality at home, this is less comfortable in countries where my marginalization is tempered by the power offered to me by neo-colonialism and the imperialism of my own culture. Traveling alone or with only other women can feel like an act of liberation and defiance, but it can also be colored with a neo-colonial sense of entitlement and a post-colonial sense of guilt.

It is perhaps worth noting at this point that the incompatibilities discussed in this chapter are particularly discomfiting because one's right to leisure abroad is not obvious or absolute. The challenges I have raised here relating to my entitlement to public space are not, for example, comparable to my entitlement to be free from sexual violence. The difference is one of degrees and also one of nature The right to exist in public space must be defended because it represents and contributes to the right to participate public life (Boys 1984). It is therefore important that women have access to the public space of their own societies equal to men. However, it is not equally, or even comparably important, that I have access to another society's public space. As a visitor deciding whether or not to transgress local norms I must consider that I might not be best placed to form a judgement on the gender dynamics of host culture based on my understanding of observable behaviors or on that of other Western visitors.

I believe the challenge facing feminist travelers is important because it is, to say the least, difficult to solve. There is at present a profound cultural clash in attitudes to gender, sexuality, and the public sphere among some of the world's cultures. Those of us who feel drawn to visit distant cultures will find this challenge difficult to manage with integrity. My only conclusion here is that we should continue to carefully consider our presence in other cultures and countries, and recognize that our wanting to go somewhere is not sufficient reason to go. That said, there is ample evidence that tourism perused conscientiously has important developmental impacts, not least on women (Nassani et al. 2019; Sheikhi et al. 2020; Zhang & Zhang 2020) and there are ethical reasons to explore the world in which we live. An addendum here is that cultural relativism does not trump the rights of women. Women's rights are human rights and many practices all over the world are, in fact, indefensible despite attempts to defend them on the grounds of cultural legitimacy. It is not necessarily within the gift of leisure tourism to make helpful stands against such oppression. However, it is certainly the obligation of us all to do what we can to ensure that women are in a position to resist their own oppression and to transgress the unjust limits which their society imposes. One way in which this aim may be furthered is through the decolonization of travel advice so that it better reflects what is genuinely considered to be appropriate by the host culture. There is significant value in offering advice to Western women traveling to post-colonial destinations from Western women who have been there themselves, however local voices are a notable absence in mainstream travel guides. Advice should be sought from local marginalized groups to provide the opportunity for visitors to understand instances where solidarity is welcome (e.g., women's initiatives which actively seek international visitors) as well as instances where transgressive behavior would be damaging. The invitation to all visitors should come from the host, and incorporating advice based on what the local population want from tourism can ensure that the information received by visitors is not simply focused on how to circumvent cultural barriers erected against them.

REFERENCES

Agovino, M., Casaccia, M., Garofalo, A. & Marchesano, K., 2017. Tourism and Disability in Italy. Limits and Opportunities. *Tourism Management Perspectives*, 23, 58–67.

Aitchison, C., 2001. Theorizing Other Discourses of Tourism, Gender and Culture: Can the Subaltern Speak (in Tourism)? *Tourist Studies*, 133–147.

Arribas-Ayllon, M. & Walkerdine, V., 2019. Foucauldian Discourse Analysis. In: C. Willig & W. Stainton Rogers, eds. *The Sage Handbook of Qualitative Research in Psychology*, 2nd Edition. London: Sage, 110–123.

BBC News, 2015. *Mount Kinabalu Naked Photo Accused Jailed.* [Online] Available at: https://www.bbc.co.uk/news/uk-england-derbyshire-33105999 [Accessed 30 August 2021].

Benjamin, S. & Dillette, A. K., 2021. Black Travel Movement: Systemic Racism Informing Tourism. *Annals of Tourism Research,* 88.

Boys, J., 1984. Women and Public Space. In: J. Boys et al., eds. *Making Space: Women and the Man-Made Environment.* London: Pluto Press.

Brown, L., 2013. Tourism: A Catalyst for Existential Authenticity. *Annals of Tourism Research*, 40, 176–190.

Brown, L. & Osman, H., 2017. The Female Tourist Experience in Egypt as an Islamic Destination. *Annals of Tourism Research,* 63, 12–22.

Córdova, T., 1998. Anti-colonial Chicana Feminism. *New Political Science*, 20(4), 379–397.

Cresswell, T., 1996. *In Place / Out of Place: Geography, Ideology and Transgression.* Minneapolis: University of Minnesota Press.

Dann, G. M. S., 2003. Noticing Notices: Tourism to Order. *Annals of Tourism Research*, 30(2), 465–484.

Elsrud, T., 2001. Risk Creation in Travelling: Backpacker Adventure Narration. *Annals of Tourism Research*, 28(3), 597–617.

Falconer, E., 2011. Risk, Excitement and Emotional Conflict in Women's Travel Narratives. *Recreation and Society in Africa, Asia and Latin America*, 1(1), 65–89.

———. 2016. "Learning to be Zen": Women Travellers and the Imperative to Be Happy. *The Journal of Gender Studies*, 26(1), 56–65.

Gavison, R., 1992–1993. Feminism and the Public/Private Distinction. *Stanford Law Review*, 45, 1–45.

Greenwood, D. J., 1989. Culture by the Pound: An Anthropological Perspective on Tourism as Cultural Commodification. In: V. L. Smith, ed. *Hosts and Guests: The Anthropology of Tourism.* Philadelphia: University of Pennsylvania Press, 171–186.

Hottola, P., 2013. Real and Imagined Women: Goddess America Meets the World. In: O. Moufakkir & Y. Reisinger, eds. *The Host Gaze in Global Tourism.* Oxford: CABI, 219–233.

Huggan, G., 1997. The Neocolonialism of Postcolonialism: A Cautionary Note. *Links & Letters*, 4, 19–24.

Jeffrey, H. L. & Stephens, S., 2021. The Subversion of Women's Anger in Travel Guidebooks. *Annals of Leisure Research.*

Kolar, T. & Zabkar, V., 2010. A Consumer-Based Model of Authenticity: An Oxymoron or the Foundation of Cultural Heritage Marketing. *Tourism Management*, 31(5), 652–664.

Lee, K. J. & Scott, D., 2017. Racial Discrimination and African Americans' Travel Behaviour: The Utility of Habitus and Vignette Technique. *Journal of Travel Research*, 56(3), 381–392.

Lonely Planet, 2021a. *Argentina in Detail: Women Travellers.* [Online] Available at: https://www.Lonely Planet.com/argentina/narratives/practical-information/directory/women-travellers. [Accessed 1 September 2021].

Lonely Planet, 2021b. *Bulgaria in Detail: Women Travellers.* [Online] Available at: https://www.Lonely Planet.com/bulgaria/narratives/practical-information/directory/women-travellers. [Accessed 1 September 2021].

Lonely Planet, 2021c. *Cuba in Detail: Women Travellers.* [Online] Available at: https://www.Lonely Planet.com/cuba/narratives/practical-information/directory/women-travellers. [Accessed 30 August 2021].

Lonely Planet, 2021d. *Egypt in Detail: Women Travellers.* [Online] Available at: https://www.Lonely Planet.com/a/nar/2f1a3372-6dbb-4ab0–91c9-61162138a161/355224. [Accessed 30 August 2021].

Lonely Planet, 2021e. *Iran in Detail: Women Travellers.* [Online] Available at: https://www.Lonely Planet.com/iran/narratives/practical-information/directory/women-travellers. [Accessed 1 September 2021].

Lonely Planet, 2021f. *Jordan in Detail: Women Travellers.* [Online] Available at: https://www.Lonely Planet.com/jordan/narratives/practical-information/directory/women-travellers. [Accessed 1 September 2021].

Lonely Planet, 2021g. *Morocco in Detail: Women Travellers.* [Online] Available at: https://www.Lonely Planet.com/morocco/narratives/practical-information/directory/women-travellers. [Accessed 1 September 2021].

Lonely Planet, 2021h. *Russia in Detail: Women Travellers.* [Online] Available at: https://www.Lonely Planet.com/russia/narratives/practical-information/directory/women-travellers. [Accessed 1 September 2021].

Lonely Planet, 2021i. *Saudi Arabia in Detail: Women Travellers.* [Online] Available at: https://www.Lonely Planet.com/saudi-arabia/narratives/practicalinformation/directory/women-travellers. [Accessed 30 August 2021].

McElroy, J.L., Tarlow,& Carlisle, K., 2008. Tourist Harassment and Responses. In: A. Woodside & D. Martin, eds. *Tourism Management: Analysis, Behaviour and Strategy.* Wallingford: CABI, 94–106.

Mehmetoglu, M., 2004. Tourist or Traveller? A Typology Approach. *Tourism Review,* 59(3), 33–39.

Mendoza, B., 2016. "Coloniality of Gender and Power: From Postcoloniality to Decoloniality." In: *The Oxford Handbook of Feminist Theory*, L. Disch and M. Hawkesworth, eds. New York: Oxford University Press, 100–121.

Nash, D., 1989. Tourism as a Form of Imperialism. In: V. L. Smith, ed. *Hosts and Guests: An Anthropology of Tourism.* Philadelphia: University of Pennsylvania Press, 37–52.

Nassani, A. A. et al. 2019. The Impact of Tourism and Finance on Women Empowerment. *Journal of Policy Modelling*, 41, 234–254.

Ong, F., Vorobjovas-Pinta, O. & Lewis, C., 2020. LGBTIQ Identities in Tourism and Leisure Research: A Systematic Qualitative Literature Review. *Journal of Sustainable Tourism.*

Paulauskaite, D., Powell, R., Coca-Stefaniak, J. A. & Morrison, A. M., 2017. Living like a local: Authentic tourism experiences and the sharing economy. *International Journal of Tourism Research*, 19, 619–628.

Pritchard, A., 2014. Gender and Feminist Perspectives in Tourism Research. In: A. A. Lew, C. M. Hall & A. M. Wiilliams, eds. *The Wiley Blackwell Companion to Tourism.* Oxford: Wiley, 314–324.

Prtichard, A. & Morgan, N. J., 2000. Privileging the Male Gaze: Gendered Tourism Landscapes. *Annals of Tourism Research*, 27(4), 884–905.

Rough Guide, 2021. *Culture and Etiquette in Morocco.* [Online] Available at: https://www.roughguides.com/morocco/culture-etiquette/. [Accessed 31 August 2021].

Sarmela, M., 1977. What is Cultural Imperialism?. In: *Cultural Imperialism and Cultural Identity.* Helsinki: Transactions of the Finnish Anthropological Society 2, 13–36.

Sheikhi, A., Ebrahimian, S. & Vazin, N., 2020. The Role of Ethnic Tourism Development on Gender Equality (A Case Study of Baluchis Women's Experience in Sistan and Baluchistan Province, Iran). *Tourism of Culture*, 1(2), 47–54.

Small, J. & Harris, C., 2012. Obesity and Tourism: Rights and Responsibilities. *Annals of Tourism Research*, 39(2), 686–707.

Stephens, S., 2020. An Autoethnography of Respectful Tourism: The Double-Bind of a Female Traveller in Morocco.. In: P. Vizcaino-Suárez, H. Jeffrey & C. Eger, eds. *Tourism and Gender-Based Violence: Challenging Inequalities.* Wallingford: CABI, 128–139.

Sunseri, L. 2000. Moving Beyond the Feminism Versus Nationalism Dichotomy. *Canadian Woman Studies*, 20(2), 143–148.

Valentine, G., 1989. The Geography of Women's Fear. *Area,* 21(4), 385–390.

Wejesingh, S. N., Mura, P. M. & Bouchon, F., 2019. Tourism Knowledge and Neocolonialism A Systematic Critical Review of the Literature. *Current Issues in Tourism*, 22(11), 1263–1279.

Wilson, E. & Little, D. E., 2005. A "Relative Escape"? The Impact of Constraints on Women Who Travel Solo. *Tourism Review International*, 9, 155–175.

———. 2008. The Solo Female Travel Experience: Exploring the "Geography of Women's Fear." *Current Issues in Tourism*, 11(2), 167–186.

Wodak, R., 2011. Critical Linguistics and Critical Discourse Analysis. In: J. Zienkowski, J. Östman & J. Verschueren, eds. *Discursive Pragmatics Vol.8.* Philadelphia: John Benjamins Publishing Company, 50–70.

Wong, M. H., 2017. *Tourists Deported after Taking Nude Pics at Angkor.* [Online] Available at: https://edition.cnn.com/travel/article/cambodia-angkor-nude-tourists/index.html. [Accessed 30 August 2021].

Yeoman, I., Brass, D. & McMahon-Beattie, U., 2007. Current Issue in Tourism: The Authentic Tourist. *Tourism Management*, 28, 1128–1138.

Zhang, J. & Zhang, Y., 2020. Tourism and Gender Equality: An Asian Perspective. *Annals of Tourism Research*, 85.

Chapter 2

(Re)Shaping the Volunteer Tourist Bubble

The Intersectional Experiences of Two Volunteer Tourist Women in Guatemala

Amy Kipp

VOLUNTEER TOURISM: AN INTERSECTIONAL FEMINIST AND ANTI-COLONIAL ANALYSIS

Volunteer tourism, defined as an alternative form of travel in which tourists pay to engage in volunteer work while traveling abroad, is a gendered sector with women making up an estimated 70 percent of volunteers (Wearing et al. 2018a). There are many models of volunteer tourism, which include both for-profit and not-for-profit trips, and a myriad of volunteer and leisure activities (Godfrey et al. 2020; Schwarz 2018; Sin et al. 2015). For example, volunteer tourists travel abroad to build wells, provide care in orphanages, teach English in schools, among many other forms of care (Kipp et al. 2020). Additionally, although instances of South-South volunteer tourism are on the rise (Baillie Smith 2018), volunteer tourists are largely young, White women from the Global North volunteering and traveling in the formerly colonized spaces of the Global South (Henry 2020; Wearing et al. 2018a). Despite the apparent feminization and Whiteness of volunteer tourism, until recently, research on the ways gender and race intersect and impact the experiences of volunteer tourists has been largely absent from the literature (Kipp et al. 2020; Bandyopadhyay and Patil 2017).

Responding to this gap and calls for more feminist research that explores the intersections of gender, race, class, sexuality, and the ongoing legacy of colonialism in volunteer tourism (e.g., Almela and Calvet 2021; Kipp et al. 2020; Wearing et al. 2018a; Bandyopadhyay and Patil 2017), in this chapter I tell the stories of Sarah, a White Canadian woman, and Luz, a Queer Latinx Woman of Color, who traveled as volunteer tourists to Guatemala in the summer of 2016.[1] There were 11 other participants on their two-week long trip to Guatemala, myself included as a participant–researcher; however, I focus on the stories of Sarah and Luz to illustrate the gendered and racialized experiences of these two women who had incredibly different experiences of the same trip. To provide context to their stories, I begin by highlighting existing feminist and other critical literature on volunteer tourism, specifically outlining the opportunities (e.g., its transformational potential), and challenges (e.g., perpetuation of a North-South divide) of volunteer tourism and the ways in which it extends the "tourist bubble" in gendered and racialized ways.[2] I share an in-depth description of the case study, including details about volunteer tourism in Guatemala and the methods I used to collect and analyze data on volunteer tourists' experiences. To illustrate the findings of this study, and center the lived experiences of women in tourism, I tell the stories of Sarah and Luz in third-person narrative.

I use these stories to foreground volunteer tourists' intersecting and frequently shifting subjectivities (e.g., gendered and racialized identities, ethnicities, sexual orientations, etc.), as well as the power structures within which these subjectivities are embedded (e.g., colonialism, heteropatriarchy, White supremacy, etc.). I argue that volunteer tourists' subjectivities (re) shape the "volunteer tourist bubble"—a concept I expand upon below—and that this (re)shaping makes the volunteer tourist bubble more or less 1) safe, 2) permeable, and/or 3) transformative for volunteer tourists as they experience different spaces, activities, and interactions while traveling in the Global South (figure 2.1). I conceptualize the volunteer tourist bubble using an intersectional feminist and an anti-colonial approach, which emphasize the interdependence and interconnection of various forms of oppression related to gender, race, ethnicity, class, disability, and other categories of difference in the context of colonialism (e.g., Pritchard 2015; Dei & Asgharzadeh 2001). I highlight the importance of thinking intersectionally about volunteer tourists' subjectivities, including the diverse experiences of those beyond the imagined "White volunteer tourist woman," as one way of prompting meaningful change within the sector. Such an approach challenges the view of the volunteer tourist bubble as a static space that is "safe" for all volunteer tourists and raises questions about how volunteer tourism might be imagined otherwise.

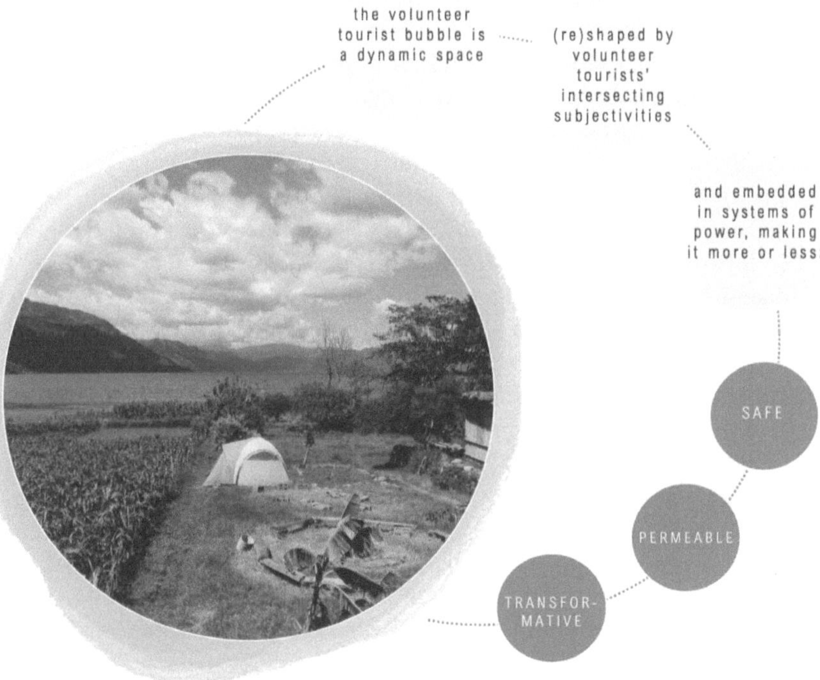

Figure 2.1. A conceptual framework of the "volunteer tourist bubble."
Source: Figure by Amy Kipp.

A "DIFFERENT" FORM OF TOURISM?

Volunteer tourism has traditionally been described as a form of sustainable alternative tourism and development that creates shared benefits to local communities and volunteer tourists (Raymond and Hall 2008; McGehee 2002; Wearing 2001). The purported benefits for communities include economic and political empowerment through the material supplies, human capital, and financial incentives provided by organizations and volunteer tourists (Wearing 2001). The sector has also been lauded for enhancing opportunities for cross-cultural connections as volunteer tourists and local communities encounter each other in increasingly intimate ways and spaces, such as caring for children in orphanages and lodging with local homestay families (Macdonald 2019; Mostafanezahd 2014; Grabowski 2013; Sin 2009). Through access to more "authentic" experiences, volunteer tourism is alleged to facilitate more genuine exchange between volunteer tourists and local communities, with the potential to transform volunteer tourists' lifestyles, values, and awareness of self (Magrizos et al. 2021; Crossley 2019; Coghlan and Gooch 2011; McIntosh and Zahra 2007) and disrupt neo-colonial tendencies perpetuated by other forms of tourism and development interventions (Wearing and Grabowski 2011; McIntosh and Zahra 2007). In this way, volunteer tourism has been positioned as one way to transform, or "burst," the tourist bubble (Mostafanezhad 2014).

Over the last decade, however, critical work has increasingly challenged this positive view of volunteer tourism. For example, Guttentag (2009) outlined negative consequences of volunteer tourism, including actions that hinder local development, promote dependency on the Global North, and facilitate unwanted cultural change. Critics have also highlighted the ongoing legacy of colonialism and neoliberal individualization embedded within the sector (Godfrey 2020; Wearing et al. 2018a; Bandyopadhyay and Patil 2017; Cheung Judge 2016). According to Jane Godfrey and colleagues (2020, 2557) volunteer tourism exists within a long tradition of "rich, White, well-educated, young people from developed countries travel[ling] to 'help' poor, usually non-White people in developing countries." Research has shown that experiences volunteering abroad often perpetuate discourses of local communities as "poor but happy," rather than disrupt volunteers' understandings of global politics and their own positions within these systems (Crossley 2012). These discourses simplify the complex nature of the economic and political relationships between the Global North and Global South (Crossley 2012; Butcher and Smith 2010; Simpson 2004) and make the practice of development seem "doable, knowable and accessible" to untrained volunteer tourists (Simpson, 2004: 681). Thus, volunteer tourism constructs volunteer tourists as the "doers" of development—or, put differently, as the "White saviours" with the power to help those "in need of saving" while traveling to "exotic" destinations in the Global South (Wearing et al. 2018a; Lahiri-Dutt 2017).

These problematic representations, which create and maintain an artificial divide between volunteer tourists and local communities, play out in practice during volunteer tourism trips in several ways. For example, Sarah Becklake (2020) describes how in Antigua, Guatemala, non-governmental organizations (NGOs) contribute to this artificial divide as they create "secure" spaces for volunteer tourists through NGO mediated project sites. Becklake (2020, 43) refers to these spaces as "safe mini bubbles" which provide volunteer tourists with a sense of security as they endeavor to go beyond what they perceive as the "inauthentic" tourist bubble. In a similar vein, Godfrey et al. (2015) found that the volunteer tourism sector in Cusco, Peru has created a "volunteer tourism enclave" in the city (141). In this enclave, the local community caters to volunteer tourists' desire for the familiar through a well-established infrastructure of backpacker bars, hostels, and adventure tourism options (Godfrey et al. 2015). Through such enclaves—which I conceptualize as the "*volunteer* tourist bubble"—volunteers can exist "separate from, but parallel to, the local community" (Godfrey et al. 2015, 141). Therefore, rather than escaping the "'tourist bubble' through intimate encounters with local people" and disrupting unequal power dynamics through affective relationships (Mostafanezhad 2014, 2), the volunteer tourist gaze is perpetuated within this expanded bubble (Godfrey et al. 2020; 2015). This gaze has been defined as the way volunteer tourists gaze upon local communities, positioning them as "Others" and furthering unequal power dynamics between the Global North and Global South (Godfrey et al. 2020).

A GENDERED AND RACIALIZED VOLUNTEER TOURIST "BUBBLE"

As stated, only recently has research on volunteer tourism begun to explore the gendered and racialized processes of volunteer tourism—which, I suggest, work to shape the volunteer tourist bubble in particular ways. For example, as a feminized sector volunteer tourism often caters to White women's problematic desires to care for racialized "Others" (Bandyopadhyay and Patil 2017; Mostafanezhad 2013). This desire is linked to current gendered manifestations of colonialism, such as romanticization about people and cultures from the Global South and discourses about women in the Global South needing rescue (Bandyopadhyay and Patil 2017). Mary Mostafanezhad (2013) argues that such discourses are reinforced by celebrity women like Angelina Jolie and Madonna who inspire young, White women from the Global North to volunteer abroad. Mostafanezhad (2013) suggests that this celebrity humanitarianism and an emphasis on "gendered generosity" in volunteer tourism works to invisibilize the unequal power dynamics at play within these interactions. Similarly, Katie Macdonald (2019) highlights how the intimate spaces of homestays can also minimize the power at play within the sector, specifically the economic, gendered, and racialized nature of these encounters. For example, through her study of volunteer tourism in Nicaragua, Macdonald (2019) showed that as volunteer women felt "adopted" into their homestay families, specifically through the care provided by their homestay "mothers," they perceived their homestays as spaces safe from the "dangerous" and "sexualized" Nicaraguan streets (Macdonald 2019). Macdonald (2019, 48) troubles the safe/homestay and unsafe/street distinction by focusing on everyday intimate encounters, arguing that "both spaces are rife with ideas and discourses about bodies, relationships, and inequalities."

The volunteer tourist bubble is also racialized in particular (i.e., White) ways. In exploring "American Whiteness in transit," for example, Jacob Henry

(2020) examines the experiences of White volunteers from the United States of America who traveled to Namibia to volunteer as teachers (Henry 2020). By analyzing their blog posts, Henry (2020, 6) found that while volunteering in the predominately "non-White spaces" of Namibia, volunteers became increasingly cognizant of their Whiteness. Although such an awareness can be a powerful tool for transforming practices of White privilege, in this study it led volunteers to "innovate ways to reassert the norm of White supremacy" while abroad, for example positioning White volunteers as celebrities or relocating their "WhiteWorlds" to White spaces in Namibia (Henry 2020, 5–6). This was also evident in a study on volunteer tourism in Honduras, in which White volunteer tourists became more aware of their Whiteness, and the privilege it afforded them, as they encountered "Others" while volunteering (Schneider 2018). Despite a desire to experience Honduras in an "authentic" way, as volunteers experienced discomfort with their White visibility/privilege they often sought out "White spaces" such as bars, coffee shops, NGO campuses, and homes frequented by other White travelers (Schneider 2018). These White spaces were spaces where volunteers could avoid the gaze of those they perceived as "Other" (i.e., local communities) and again feel comfortable with their Whiteness (Henry 2020; Schneider 2018).

Although much of the research exploring gender and race in volunteer tourism examines the experiences of White women, many people beyond the "archetypal young White elite" are volunteering abroad (Cheung Judge 2016, 239). For instance, in her work on embodied performances of race during a trip in Zimbabwe, Ruth Cheung Judge (2016) explored the experiences of young Black British volunteer tourists and how they negotiated their Blackness while in Zimbabwe. In this study, "race materialized in multiple and complex ways" for these volunteers (244). Black volunteers often expressed a sense of dissonance with the "racialized hierarchies of charity" (239) (i.e., White as savior, Black as "poor") while simultaneously mobilizing their Blackness as the basis for their felt connection to Africa (Cheung Judge 2016). These racialized experiences were embodied and affective. Enacting social practices and preferences around eating, such as responses of delight or disgust to local foods and eating specific meals without utensils, were common ways Black volunteer tourists negotiated their identification with the continent. Cheung Judge (2016) found that Black volunteer tourists performed their Blackness in an agentive way. However, their agency was bounded by existing power dynamics such as the legacy of colonialism and racism that underpin volunteer tourism.

In this chapter, I further conceptualizations of the volunteer tourist bubble by building on Smith's (1978) concept of the "tourist bubble" and recent literature that has described the "volunteer tourist bubble" using terms such as the "volunteer tourist enclave" (Godfrey et al. 2020; 2015), "White space"

(Schneider 2018), "WhiteWorld" (Henry 2020), and "safe mini bubbles" (Becklake 2020). I suggest that existing conceptualizations have largely focused on the boundaries organizations and volunteers establish between themselves and host communities, conjuring images of a protective wall. However, bubbles are also characterized by their fragility, non-fixity, and non-uniformity. It is these characteristics that I evoke throughout this chapter to further expand the concept of the volunteer tourist bubble. I add to the existing literature on the volunteer tourist bubble (Godfrey et al. 2020, 2015; Becklake 2020; Henry 2020; Schneider 2018) and the gendered and racialized experiences of volunteer tourists (Wearing et al. 2018a; Bandyopadhyay and Patil 2017; Mostafanezhad 2013; Cheung Judge 2016) by exploring how volunteer tourists' gendered, racialized, and other subjectivities intersect to influence their experiences within this bubble and how these subjectivities, in turn, (re)shape the bubble in distinct ways. I argue that in order to imagine sustainable and regenerative alternatives to volunteer tourism, a more nuanced understanding of the volunteer tourist bubble is needed, one which uses an intersectional approach to consider the diverse experiences beyond the archetypal "White woman volunteer tourist."

EXPLORING A VOLUNTEER TOURIST BUBBLE IN GUATEMALA: A CASE STUDY

This chapter presents a case study of a two-week long volunteer trip to Guatemala. This case study is a part of a larger project that explored the gendered and racialized experiences of volunteer tourists traveling and volunteering to the Global South, which I conducted for my master's research in 2016. In addition to this case study, this project included thirty interviews with individuals who had traveled abroad to participate in development focused volunteer projects (see Kipp et al. 2020). The purpose of conducting a case study was to build on this set of interviews to further understand the embodied and affective experiences of negotiating gendered and racialized subjectivities abroad through participant observation.

For over a decade, Guatemala has been a popular destination for individuals seeking to volunteer abroad (see Riddering 2020; Becklake 2020, 2014; Vodopivec and Jaffe 2011). In 2011, Vodopivec and Jaffe detailed the presence of volunteer tourists throughout the country, including: "[volunteers] working with sea turtles on the Pacific Coast or on archaeological sites in El Petén, building houses in the highlands of Central Guatemala, or helping out in women's shelters in Guatemala City" (116). There is an abundance of NGOs in Guatemala that facilitate international volunteers looking for both a way to "help" and for a "safe" entry point into the country (Riddering 2020;

Becklake 2020, 2014). In Guatemala NGOs, which have increasingly incorporated volunteer tourism into their financial and operational models, often emphasize poverty and violence in the country in their promotional materials to recruit volunteer tourists (Becklake 2020). Although the idea of "risky Guatemala" may cause some travelers to avoid the country, Becklake (2020) suggests that this imaginary appeals to volunteer tourists interested in traveling off the beaten path. However, within this context travelers are also drawn to the security provided by an NGO, which arranges their accommodations, volunteer placements, and leisure activities (Becklake 2020). Additionally, as they attempt to inspire compassion in potential volunteer tourists NGOs often reproduce the "humanitarian gaze," positioning Guatemalans as individuals in need of "saving" (Becklake 2020; Mostafanezhad 2013). At the time of this research, I had traveled to Guatemala during two short-term volunteer tourism trips (in 2010 and 2011) as well as a three-month long semester abroad (in 2013). These visits provided me with a preliminary understanding of the cultural, political, and historical context of Guatemala as well as insight into existing practices of volunteer tourism in the country.

The volunteer trip at the center of this case study was organized by a Canadian NGO, which I refer to as *Purposeful Travel.*[3] *Purposeful Travel* has had operations in Guatemala for over a decade. Differing from more "traditional" volunteer tourism organizations, *Purposeful Travel* focuses on solidarity with and learning from local partners while traveling and volunteering. The trip to Guatemala was designed for volunteer tourists to learn about Mayan culture, social justice, and environmental activism. While abroad, volunteer tourists participated in reforestation and a local permaculture project through tree planting and other agricultural and construction tasks. We also traveled to three destinations throughout the country, staying in hostels, with homestay families, and at the organizations' headquarters. *Purposeful Travel* promised us that we "would live like the locals" and "get [our] hands dirty" as we traveled with a group of like-minded backpackers throughout Guatemala.

The group—which participants often referred to as "the team"—was comprised of 13 individuals who were recruited and selected by *Purposeful Travel.* We were 10 volunteers, two team leaders, and myself as a participant-researcher, all between the ages of 18 and 30 and citizens of either the United States or Canada. Eleven team members identified as women, and two as men. Of this group, three women identified as Queer. The majority (eight members) identified as White. One woman identified as a Vietnamese-American, and one woman as Queer and Latinx, specifically El Salvadorian and Guatemalan (Luz). Luz was also fluent in Spanish, as were the two team leaders. The rest of the team spoke little to no Spanish.

As researcher, I took part in the same activities as the other volunteers: I shared hostel rooms and meals. I spent time with our local partners and homestay families. I planted trees and carried buckets of water, and I sat through orientation and debrief sessions, among other shared moments. I was (often uncomfortably) positioned as an "insider" with the team, which enabled me to build trust and rapport and led to an "in-depth exploration of the volunteer tourism encounter" (Sin 2009, 487). However, this insider position was also incredibly challenging. As problematic interactions occurred—which I detail in this chapter—I struggled navigating my various subjectivities as a researcher and participant, as a White woman traveling and volunteering in Guatemala, and as someone committed to feminist and anti-colonial praxis (Faria and Mollett 2016; Andrews and Gupta 2010).

I conducted conversational style interviews upon returning home from Guatemala. Ten team members chose to participate in these follow-up interviews to debrief on their/our experiences abroad. I focus here on the stories of two women from this case study, Sarah and Luz, which I share in third-person narrative form as a way of doing academic "writing differently" (Mandalaki and Daou 2021; van Eck et al. 2021). This narrative approach enables an embodied and affective account of the women's negotiations of their subjectivities as volunteer tourists in Guatemala. Although I have elected to share stories of two very different experiences, the themes I expand on through my discussion are consistent with the findings from this specific case study as well as those of the broader study.

A WHITE WOMAN'S "LIFE CHANGING" EXPERIENCE OF VOLUNTEER TOURISM IN GUATEMALA

Identifying as a cisgender, White woman from the East Coast of Canada, Sarah worked as a tradesperson in the Canadian prairies. She grew up in a community, which she described as "not really having any other races" aside from White folks. Sarah was motivated to volunteer abroad after seeing a YouTube video promoting *Purposeful Travel*. She chose a trip to Guatemala from the organization's website because it "lined up perfectly with the days off" she already had scheduled. Sarah had recently gone through "a pretty bad break up" and wanted to travel but did not think it would be safe for her to go alone. She had never traveled to a country in the Global South before. "I felt like 'Eat, Pray, Love'" she said, "I felt like Julia Roberts, like I'm going to go find myself. I really thought it was going to be a life changing experience, and it was." She liked the idea that through volunteering she could "get in with the locals." She was looking for an experience that would be more "uplifting" than simply "drinking, partying, and travelling." The purpose of the trip [was] not just volunteering . . . It [was] to enlighten [volunteers] about what's going on [in the world]."

Describing her time in Guatemala, Sarah explained, "It was a really crazy experience for me. And, I found that it was really eye opening and it really changed me." Traveling and volunteering in Guatemala shifted Sarah's understandings about the day to day lives of families experiencing poverty. She explained, "I [used to] think that if you're poverty stricken there's some type of like depression or sadness that [would come with that] but I felt like people there weren't like that at all. They're so happy and they just don't see [their poverty] as a misfortune."

Deeply impacted by an afternoon the group spent with a local activist named Martin[4], Sarah described learning more about "White privilege" while in Guatemala. Martin worked with others to imagine—and build—a Guatemala in which Guatemalans have the opportunities and resources they need to stay in the country, rather than migrating to the United States. Sarah explained, "I definitely would never [have] consider[ed] myself racist, but I think that I had some preconceived notions of [immigration]. [Before], I would have been like 'why don't they just stay in their country?' . . . I couldn't really see that there was a need . . . It was poverty and a necessity to feed their family [that led to migration]." Sarah asserted, "I didn't really get that and now I definitely get that."

Traveling with a group of mostly women was important. As a woman working in the trades, Sarah's work is male dominated, and she was eager to bond with the other volunteer women. The "girl power" and solidarity that came from volunteering with other women motivated Sarah during the challenging parts of the trip. "If [a fellow volunteer woman] can do it, I can do it!"

Although Sarah valued her time in Guatemala—including the lessons she learned while abroad and the people she met—she did not want to go on another volunteer trip in the future. She struggled with the lack of control she had, including where she traveled, what she ate, and what she wore. For example, she felt that no one took her food preferences seriously and that since she was "paying to be fed," she should be able to "eat what she wanted to eat." After a particularly grueling hike, she declared "if there are beans for dinner, I think I will cry" and, as the Guatemalan hosts served rice and beans, she refused the meal.

Sarah also struggled with the dress code outlined by the organization, which required volunteer tourists to dress "modestly" while in certain areas. Frustrated by these rules and being asked by a program facilitator to "cover up" while using a traditional Mayan sauna, Sarah announced to no one in particular, "I'm getting fucking sick of being in a country where I can't be a woman. I have tits, get over it." Sarah felt like the group leaders "put way more stress on [the dress code] then there needed to be" and she wondered if Guatemalans even cared about the way she dressed. She "guaranteed" that if she showed up wearing a tank top and shorts at the tree planting site Paco, a local community organizer, would just be grateful she was working regardless of what she was wearing. Sarah questioned whether "it was an [organization] issue" rather than a "Guatemalan people issue." "Maybe they [the Guatemalan organizers] did have issues but they didn't say it. And I didn't feel like we were being disrespectful."

Before volunteering in Guatemala, Sarah felt that as a woman, traveling abroad was safer in a group. Her family would have "lost their minds" if she tried to travel anywhere by herself, or even with a friend. Although not afraid of traveling solo, she believed people would try and take advantage of a woman traveling alone. After spending time in Guatemala, Sarah believed she would now be able to do the next trip by herself because she had "learned the ropes"—but, she would "obviously research it a lot and make sure that it is somewhat safe for women."

Returning home after being in Guatemala was hard for Sarah, she described "reverse culture shock" and told of her frustration when she spoke to her friends and family about her trip. "They were being very normal . . . and I would just break down and start crying and say like 'you don't even get it, like you don't even care.'" She struggled with feeling "completely different" while everyone else had "just gone to work for two weeks." People would "complain about something so menial," Sarah would want to scream, "NO! YOU DON'T UNDERSTAND. YOU DON'T KNOW WHAT THEY'RE DEALING WITH IN GUATEMALA!"

In an effort to cope with these intense emotions, Sarah reached out to her cousin who had "gone to Africa" to volunteer. Her cousin had experienced similar emotions after returning to Canada and reassured Sarah, "you'll get into your normal routine again, things will feel normal." Sarah agreed. Although she felt like "a mess" for the first week she was home, after she went back to work and time passed, she "was fine."

A QUEER LATINX WOMAN'S "OPPRESSIVE" EXPERIENCE OF VOLUNTEER TOURISM IN GUATEMALA

Luz identified as Guatemalan, El Salvadorian, Latinx, and a Queer Woman of Color. She was born in Toronto, Canada, and acknowledged the privilege that came with "having a Canadian passport" but did not identify as Canadian. Luz decided to volunteer in Guatemala as a way to "enquire more about [her] background and [her] roots." She did not currently talk to the Guatemalan side of her family and so had no "means of learning about or asking someone about Guatemala or the culture." It was "important for [her] as a woman to figure it out by [herself]." She had traveled to El Salvador many times in the past and had been to Guatemala once before her volunteer trip. She researched volunteer organizations online when an advertisement for a trip to Guatemala with *Purposeful Travel* appeared on Facebook. She applied to join the trip—which claimed to be focused on Mayan culture—"expecting to learn about Mayan roots, culture [and] history."

In addition to learning about her "roots," Luz explained that she also wanted to travel abroad to contribute to women "taking up space." "If there was anything empowering about the group," Luz said, "it was there being a lot of women there. We're so used to men taking up so much more space than women . . . so it was nice being in a space [that was] predominantly women identifying." However, Luz pointed out that even in such a space "there were still people who upheld the patriarchy."

Despite her expectations and her ability to take up space, Luz's experience in Guatemala—specifically with her teammates and the program facilitators—was one she described as "problematic." For her, "it was very difficult to make it to the end of the two weeks." She explained that from "day one" she was like "here we go, a group of White folks from the States and Canada, [even] the leaders are White." This dynamic immediately put her "on the defensive." Luz elaborated on how her Latinx identity influenced her experiences with the team, "it was like the group was choosing which identity would come out for me." In spaces like hostels, where the volunteer tourists only interacted with one another, Luz often felt like they just saw her "like everyone else" but in public spaces she often felt like the group viewed her as the "token Latina."

Luz also recounted stories in which the actions, attitudes, and language of the other volunteer tourists and facilitators were "super violent in many ways in terms of racism, sexism, classism, and homophobia." For example, over the course of the trip, the other volunteer tourists had grown increasingly frustrated with being served the typical Guatemalan cuisine—specifically rice, tortillas, and beans. They were constantly complaining about the food, which Luz understood as complaining about Guatemalan culture. She reflected on one specific interaction with Sarah. One morning during the last few days of the trip, Sarah who had expressed her dissatisfaction with the food throughout, approached Luz with a box of pancake mix purchased from a local tienda. Referring to the pancake mix in hand, Sarah exclaimed to Luz, "do you know what this says? This says: 'fuck frijoles, fuck beans.'" Luz was outraged, and explained, "I was like you have literally just insulted me to the core and you don't even realize it." She remembered feeling as though she could not even "make a fuss about this" because no one would understand. She stated, "everyone is going to think it's just food, but it's much deeper than that The complaining of beans, like that's my culture. That is what people here eat, that is what I eat here." Connecting this interaction to her Latinx identity, Luz stated, "I'm just not going to get over the fact that she chose to say that to *me*."

Providing another example, Luz described a conversation the team had after learning from a local partner about the legacy of colonialism in Guatemala. In this organized debrief session, team leaders attempted to make connections between this legacy and the ongoing racism, sexism, and homophobia in Canada and the United States. In response, a White American volunteer man stated, "it's easy to point fingers at straight White men," and explained that his "gay and Black friends" often received "perks" associated with their identities. He then proceeded to make several overtly homophobic remarks. Luz was outraged, but not surprised, that this volunteer was "arguing that the system had failed a White man." She remembered thinking, "holy shit, I cannot believe this is being said right now or tolerated . . . But of course it is being tolerated, because it's a group of White people who obviously don't understand this conversation or what it feels like to be in these situations." Outraged, she left the group debrief session.

Despite having an overwhelmingly negative experience, Luz also told of key moments where she was able to relationally connect with Guatemalan partners and local community members. Luz spoke about why she stayed on the trip in light of her negative experiences, stating: "[I was thinking] about taking up as much space as everyone else is taking up and that I deserve to be here, that I have the right to be here, and no one is going to cancel me out from that. It was almost just like an act of resistance . . . I'm not going to . . . be uprooted because of these people." Luz had a conversation with Martin, a local activist, who encouraged her to stay. Martin stated, "your ancestors walked these streets, regardless of if it's Guatemala or any other Latin American country this is where you are from and don't let other people . . . affect your travels."

In addition to her connection with Martin, Luz developed meaningful relationships with other Guatemalan partners and local community members. She spoke about her connections in the context of the homestay family, explaining: "It was obviously very easy for me to get along because I spoke the language, so it was just natural . . . I loved the family. They were very welcoming. I took it as an opportunity to finally learn about Mayan roots, like small things like how to make tortillas or asking questions just about food like 'how do you make this, so I can make this back home' or like 'what are the steps to doing this corn tea'? It felt like I was at my Grandma's house. . . . It just felt like all these memories of the Latin countries that I have travelled to just like all in one, just so comfortable."

Although Luz was largely comfortable in "Guatemalan spaces," in which she often felt like "she was Guatemalan," her feeling of "being Guatemalan" was nuanced. She explained, "even within those spaces I was still [perceived of as] White because I'm not from there. I wasn't born there. I don't wear what they wear. I don't do what they do." For her this experience was "very . . . displacing and diasporic." Contributing to this feeling was her queer identity. She described the impact of her Queerness, explaining that there were just "little things" that she picked up that reflected a culture of machismo in Guatemala, like "asking someone if they have a boyfriend and not considering that you could not have an opposite sex relationship." Even if she "romanticizes Guatemala" and wants to move there one day, she realized she would "still have part of [her] identity oppressed."

After returning home, Luz was angry and more pessimistic than she had been before traveling as a volunteer to Guatemala. She was angry with the organization and the other volunteer tourists. From her perspective, the program facilitators should have "taken more responsibility for how problematic the program was." The organization should have had "someone who is actually a Spanish speaker, someone from Guatemala," in a leadership role. Luz also felt angry more generally with people born in Canada who are privileged and uninformed about issues of social justice. She explained that she was "learning to deal with that anger and turn it into something constructive and positive and motivating." "I mean fuck," she concluded, "our work is cut out for us."

(RE)SHAPING THE VOLUNTEER TOURIST BUBBLE

These divergent narratives trouble easy understandings of the volunteer tourist bubble as a unified space that forms a "protective wall" between volunteer tourists and local communities. I theorize the volunteer tourist bubble as a dynamic space that is (re)shaped by volunteer tourists' subjectivities and explore the volunteer tourist bubble as a more or less 1) safe, 2) permeable, and 3) transformative space. From a feminist and anti-colonial perspective, I highlight the potential challenges and opportunities of a volunteer tourist bubble with these characteristics.

THE VOLUNTEER TOURIST BUBBLE AS AN (UN)SAFE SPACE

Existing research identifies safety as a key concern for women traveling from the Global North to the Global South (e.g., Kipp et al. 2020; Wilson and Little 2008; Jordan and Gibson 2005). Although material threats often exist for women travelers (Su and Wu 2020), imaginaries of the Global South as a dangerous place can be understood within the context of colonialism and patriarchy (Cook 2007; Jordan and Gibson 2005). These structures position racialized men as dangerous and "hyper-sexual" and White women as the object of sexual desire (Cook 2007; Jordan and Gibson 2005). Sarah and her family's concern for her safety while traveling abroad can also be understood within this context. Like other volunteer tourists (Kipp et al. 2020), Sarah acknowledged that she was drawn to traveling with a volunteer tourism organization because of the perceived safety provided through the trip's set itinerary, group travel, and connections to local communities. These elements work to shape the "protective wall" of the volunteer tourist bubble, positioning some as inside its boundaries (i.e., volunteer tourists, organization staff, and local partners) and "Others" (i.e., local people not connected to the organization) as outside. Although the individuals within this particular volunteer tourist bubble were not exclusively White (i.e., there were racialized volunteer tourists and local partners), I suggest that it was their proximity to and familiarity with Whiteness that positioned them as "safe" for volunteer tourists such as Sarah.

Despite this (problematic) perception of safety, the bubble may not be perceived as a safe space for all volunteer tourists. For example, Luz's experience was negatively impacted by comments disparaging Guatemalan culture, practices that positioned Luz as the "token Latina," and the dismissal of violence experienced by the Queer community. Such discourses and practices were perpetuated by volunteer tourists and the organization, marginalizing Luz in terms of her racialized, cultural, and sexual identities. Similar to the findings of Cheung Judge (2016) in which volunteer tourists' racialized/cultural identities were performed and reinforced through their responses to local food, Luz spoke about the connection between food, her Guatemalan culture, and her experience as a volunteer tourist. Other volunteer tourists on the team continually complained about Guatemalan food and other aspects of Guatemalan culture, contributing to Western/White superiority which positioned Guatemalans, and therefore Luz, as inferior.

Purposeful Travel's group debrief sessions also contributed to Luz's negative experience of the trip. During one of the group debrief sessions a fellow volunteer, in this case a White American man, made comments that negated the structural barriers faced by Black and Queer communities. "But of course this is being tolerated by a group of White people," Luz asserted, connecting the group's tolerance of homophobia and racism to the Whiteness of the bubble. Rather than creating a sense of familiarity and safety, the bubble was a displacing and unsafe environment for Luz—and arguably other Queer People of Color.

Sarah and Luz's stories reveal how the volunteer tourist bubble is often marketed as a "safe space" for specific bodies traveling in the Global South (i.e., Western, White, straight women and men). This imagined sense of safety may contribute to the proliferation of people travelling abroad with volunteer tourism organizations—specifically women from the Global North (Kipp et al. 2020). However, imagining the volunteer tourist bubble as a safe space can also reinforce White supremacy and heteropatriarchy and lead to the exclusion and inclusion of certain people. When those other than the normative volunteer enter the bubble, they may in fact experience the bubble as an unsafe space.

THE PERMEABILITY OF THE VOLUNTEER TOURIST BUBBLE

Sarah and Luz's experiences in Guatemala also illustrate how the borders of the volunteer tourist bubble are more or less permeable based on the rules of organizations and the subjectivities of volunteer tourists. I use the term permeability to indicate the opportunities volunteer tourists had to relate with people and inhabit spaces outside of the bubble. For example, in this case study the borders of the bubble were relatively impermeable for volunteer tourists with subjectivities similar to Sarah's (i.e., Western, White women). *Purposeful Travel* largely controlled the spaces they could access, the people with whom they interacted, and the activities in which they participated.

The multiple and shifting subjectivities of participants, including their gender, ethnicity, nationality, and sexual orientation, also shape the permeability of the volunteer tourist bubble. Luz was able to make more intimate connections with Guatemalan homestay families, local partners, and people she met in markets, *tiendas*, restaurants, and other public spaces. Luz believed such connections were a direct result of her Latinx identity, cultural, and linguistic knowledge. She described this permeability as being able to easily navigate English and Spanish spaces. She used words like "very easy," "natural," and "comfortable" when speaking about her experience with her homestay family. As Luz engaged in Guatemalan practices, the borders of the bubble also became more permeable. By asking questions while making tortillas, eating, and appreciating local food, and learning to weave, Luz began to disrupt the divide between herself and local communities. However, to a certain degree Luz believed that her Queer identity, her "not being from there," and her perceived Whiteness, maintained a sense of separation between her and local Guatemalans. Even as the volunteer tourist bubble was permeated, certain barriers remained.

On the contrary, Sarah's experience with her homestay family was marked neither by comfort nor a sense of familiarity. She viewed such interactions as largely transactional (i.e., something she was "paying for") and was frustrated by a perceived lack of accommodation for her food and clothing preferences.

These moments indicate that some volunteer tourists may be more comfortable in a relatively impermeable bubble, in which they are surrounded by familiar gender, racialized, and cultural norms. These preferences may encourage organizations to create experiences that limit any form of sustained connection between volunteer tourists and local communities. For *Purposeful Travel*, this included employing team leaders from the Global North rather than local leaders, arranging stays in hostels or at NGO campuses rather than exclusively homestays, and limiting opportunities for relationship building with people other than fellow volunteers. In this way, they actually reduced opportunities for relationality and learning needed to foster solidarity between people in the Global North and Global South (Kipp et al. 2020).

Important lessons can be learned by considering what subjectivities and practices influence the permeability of the volunteer tourist bubble. Considering Sarah's experience in the bubble as relatively impermeable reveals the ways in which White comfort is maintained and prioritized even within the racialized spaces of the Global South. The volunteer tourist bubble was more permeable for Luz as she connected with community members based on shared identities and through engaging in local practices. Volunteer tourism trips that create a more permeable volunteer tourist bubble—or, even further seek to "burst" the volunteer tourist bubble entirely—can begin to challenge the power hierarchies embedded in volunteer tourism encounters. Centering local leadership and creating trips more inclusive of volunteer tourists with diverse identities are two strategies that may contribute to this goal.

THE VOLUNTEER TOURIST BUBBLE: A TRANSFORMATIVE SPACE?

The volunteer tourist bubble can be a space for both the transformation *and* reinforcement of volunteer tourists' existing ideologies, around gender, cultural norms and their understandings of global power dynamics. As illustrated in this case study the volunteer tourist bubble can enable women to "take up space" in the traditionally masculinized space of international travel (Wearing et al. 2018b). As an increasing number of women feel comfortable traveling abroad through volunteer tourism, existing understandings of tourism as a gendered (masculinized) space can also be challenged. For Sarah this challenge to the masculinized space of international travel was experienced as a feeling of "girl power" in the volunteer tourist bubble; for Luz as the only "empowering" aspect of the trip. However, there is also potential for Western understandings of gender to be reinforced in the volunteer tourist bubble. For instance, when Sarah resisted the dress code and exclaimed "I'm getting fucking sick of being in a country where I can't be a woman" she revealed that her understanding of what it means to "be a woman," based in Western gender norms such as what clothes women wear, was troubled during her time abroad. Although Sarah understood this to be a gendered restriction, team leaders explained that the dress code was actually in place to respect Mayan cultural norms of both men and women dressing modestly. This interaction illustrates how cultural misunderstandings (Raymond and Hall 2008) can be reinforced within the volunteer tourist bubble.

The transformational potential of the volunteer tourist bubble can also be examined in the context of colonialism and its ongoing legacy in tourism and development encounters (Bandyopadhyay and Patil 2017). Romanticized notions of "finding oneself" through volunteer tourism positions the Global South as the backdrop for Western volunteer tourists' transformation. Although Sarah experienced her trip abroad as "lifechanging," pointing to a deeper understanding of White privilege and factors leading to outmigration from Guatemala, her story also indicates the potential limitations of volunteer tourists' transformation. For example, her interactions with a local homestay family led Sarah to characterize Guatemalans in a way that mirrored the trope of people in the Global South as "poor but happy," invisibilizing the challenges experienced by those marginalized by capitalism and colonialism (Crossley 2012). Sarah's experience also challenges the longevity of perceived transformations that occur in the volunteer tourist bubble (see Crossley 2019). Although she struggled emotionally upon returning to Canada, she explained that she was "fine" after her life went back to "normal." Thus, although some interactions volunteer tourists have in the bubble have the potential to transform ideas about privilege and relationships between the Global North and Global South, I suggest that a bubble characterized by Whiteness and limited interactions with local communities actually restricts opportunities for sustained transformation.

As seen in Luz's narrative her presence as a Queer Latinx Woman of Color disrupted the Whiteness of the volunteer tourist bubble and highlights the nuances of its transformational potential. Her story raises questions about what bodies experience the volunteer tourist bubble as transformational and what this transformation involves. Luz described group debrief sessions as oppressive, even though they were created by *Purposeful Travel* to facilitate transformative learning for volunteer tourists. During such sessions, the norms of White supremacy and heteropatriarchy were often maintained rather than challenged.

Importantly, Luz characterized her response to her experiences in the bubble as anger which, rather than leading to inaction, caused Luz to consider how she could turn her indignation into "something constructive and positive and motivating." In this way her anger worked to reinforce her commitment to resisting oppressive power dynamics and, in an unexpected way, acted as a potential catalyst for change (see DeGagne 2018 on anger and its uses in activism). Although existing research has highlighted the disillusionment experienced by some volunteer tourists (Zavitz and Butz 2011), more is needed to examine the impact of negative experiences within the volunteer tourist bubble. Additionally, despite her negative experience with other volunteer tourists and the organization itself, for Luz it was through relational connections with Guatemalans, and "small things" like learning to make tortillas, that she learned about her Mayan roots. Perhaps it is the moments in which the volunteer tourist bubble is most permeable that create space for change.

CONCLUSION

In this chapter, I build on conceptualizations of the volunteer tourist bubble by examining the distinct experiences of two volunteer tourist women during the same trip in Guatemala. Thinking intersectionally about the identities of volunteer tourists highlights the diverse ways gender, race, ethnicity, and sexual orientation shape the bubble. Analyzing these experiences using a feminist and anti-colonial approach draws attention to the ways in which the everyday experiences of volunteer tourists (e.g., eating, cooking, dressing, speaking, etc.) are imbedded in the same power hierarchies (e.g., heteropatriarchy, colonialism, and White supremacy) that shape the bubble in traditional tourism encounters. Colonial and heteropatriarchal ideas about what spaces are safe for women from the Global North and what bodies are safe to interact with are also present in volunteer tourists' imaginaries of the Global South. And, subtle and not-so-subtle practices and discourses around White supremacy surface as White volunteers seek out familiar practices, people, and places within the volunteer tourist bubble.

The experiences of Sarah and Luz provide critical insight into the power dynamics still at play within the sector and underscore that fundamental change is needed. Changes could include organizations run by local communities and facilitated by local community members; feminist, anti-colonial, and anti-racist teaching embedded throughout volunteer tourists' experiences; and organizations intentionally recruiting and providing resources for volunteer tourists with diverse identities, including racialized people, those who are LGBTQ+, and folks with disabilities. The global history and ongoing legacy of colonialism, globalization, and heteropatriarchy both in the Global North and Global South present barriers to implementing these suggestions. However, I suggest that to (re)shape the tourist bubble, researchers and organizations must begin by critically reimagining the volunteer tourist.

REFERENCES

Adiyia, B., A. Stoffelen, B. Jennes, D. Vanneste, and W. M. Ahebwa. 2015. "Analysing Governance in Tourism Value Chains to Reshape the Tourist Bubble in Developing Countries: The Case of Cultural Tourism in Uganda." *Journal of Ecotourism* 14(2–3): 113–129.

Almela, Marta Salvador, and Núria Abellan Calvet. 2021. "Volunteer Tourism and Gender: A Feminist Research Agenda." *Tourism and Hospitality Research.*

Andrews, H. and P. Gupta. 2010. "Researching Tourism: Reflexive Practice and Gender." *Journal of Tourism Consumption and Practice* 2(2): 1–14.

Baillie Smith, M., N. Laurie, and M. Griffiths. 2018. "South–South Volunteering and Development." *The Geographical Journal* 184(2): 158–168.

Bandyopadhyay, R., and V. Patil. 2017. "'The White Woman's Burden'–the Racialized, Gendered Politics of Volunteer Tourism." *Tourism Geographies* 19(4): 644–657.

Becklake, S. 2020. "The Role of NGOs in Touristic Securitization: The Case of La Antigua Guatemala." *Space and Culture* 23(1): 34–47.

Braun, V., and V. Clarke. 2019. "Reflecting on Reflexive Thematic Analysis." *Qualitative Research in Sport, Exercise and Health* 11(4): 589–597.

Butcher, J., and P. Smith. 2010. "'Making a Difference': Volunteer Tourism and Development." *Tourism Recreation Research* 35(1): 27–36.

Carrier, J. G., and D. V. Macleod. 2005. "Bursting the Bubble: The Socio-cultural Context of Ecotourism." *Journal of the Royal Anthropological Institute* 11(2): 315–334.

Cheung Judge, R. 2016. "Negotiating Blackness: Young British Volunteers' Embodied Performances of Race as They Travel from Hackney to Zimbabwe." *Young* 24(3): 238–254.

Coghlan, A., and M. Gooch. "Applying a Transformative Learning Framework to Volunteer Tourism." *Journal Of Sustainable Tourism* 19(6): 713–728.

Crossley, É. 2012. "Poor but Happy: Volunteer Tourists' Encounters with Poverty." *Tourism Geographies* 14(2): 235–253.

———. 2019. "Volunteer Tourism: Subjectivity and Transformation." *Critical Tourism Studies Proceedings* 1(2019): 75.

DeGagne, A. (2018). "On Anger and Its Uses for Activism." *Contemporary Inequalities and Social Justice in Canada*, 142–163.

Dei, G. J. S., and A. Asgharzadeh. 2001. "The Power of Social Theory: The Anti-colonial Discursive Framework." *The Journal of Educational Thought (JET)/ Revue De La Pensée Éducative,* 297–323.

Faria, C. and S. Mollett. 2016. "Critical Feminist Reflexivity and the Politics of Whiteness in the 'Field.'" *Gender, Place & Culture* 23(1): 79–93.

Godfrey, J., S. Wearing, and N. Schulenkorf. 2015. "Neo-colonialism and the Volunteer Tourist Gaze: Commercial Volunteer Tourism in Cusco, Peru." *Current Issues in Tourism* 23(20): 2555–2571.

Godfrey, J., S.L. Wearing, N. Schulenkorf, and S. Grabowski. 2020. "The 'Volunteer Tourist Gaze': Commercial Volunteer Tourists' Interactions with, and Perceptions of, the Host Community in Cusco, Peru." *Current Issues in Tourism* 23(20): 2555–2571.

Grabowski, S. 2013. "Acculturation, Re-entry and Transformation: The Story of a Volunteer Tourist." In *Transformational Tourism: Tourist Perspectives*, chap. 14.

Guttentag, D. A. 2009. "The Possible Negative Impacts of Volunteer Tourism." *International Journal of Tourism Research* 11(6): 537–551.

Henry, J. 2020. "Whiteness in Transit: The Racialized Geographies of International Volunteering." *Social and Cultural Geography* 23(3): 1–17.

Hesse-Biber, S. 2014. "Feminist Approaches to In-depth Interviewing." In *Feminist Research Practice: A Primer*, 182–232.

Jacobsen, J. K. 2003. "The Tourist Bubble and the Europeanisation of Holiday Travel." *Journal of Tourism and Cultural Change* 1(1): 71–87.

Judd, D. R. 1999. "Constructing the Tourist Bubble." *The Tourist City*: 35–53.

Lahiri-Dutt, K. 2017. "Thinking 'Differently' About a Feminist Critical Geography of Development." *Geographical Research* 55(3): 326–331.

MacDonald, K. 2019. "Between Catcalls and Sharing Space: Intimacy in Volunteer Abroad." In *Women's Studies International Forum* 73: 42–49.

Magrizos, S., I. Kostopoulos, and L. Powers. "Volunteer Tourism as a Transformative Experience: A Mixed Methods Empirical Study." *Journal of Travel Research* 60(4): 878–895.

Mandalaki, E., and E. Daou. 2021. "(Dis) Embodied Encounters Between Art and Academic Writing Amid a Pandemic." *Gender, Work and Organization* 28: 227–242.

McGehee, N. G. 2002. "Alternative Tourism and Social Movements." *Annals of Tourism Research* 29(1): 124–143.

McIntosh, A. J., and A. Zahra. 2007. "A Cultural Encounter through Volunteer Tourism: Towards the Ideals of Sustainable Tourism?" *Journal of Sustainable Tourism* 15(5): 541–556.

Mostafanezhad, M. 2013. "'Getting in Touch with Your Inner Angelina': Celebrity Humanitarianism and the Cultural Politics of Gendered Generosity in Volunteer Tourism." *Third World Quarterly* 34(3): 485–499.

———. 2014. *Volunteer Tourism: Popular Humanitarianism in Neoliberal Times*. Surrey: Ashgate Publishing.

Pritchard, A. 2014. "Gender and Feminist Perspectives in Tourism Research." *The Wiley Blackwell Companion to Tourism,* 314–324.

Raymond, E. M., and C. M. Hall. 2008. "The Development of Cross-Cultural (Mis) Understanding Through Volunteer Tourism." *Journal if Sustainable Tourism* 16(5): 530–543.

Schneider, M. J. 2018. "Exotic Place, White Space: Racialized Volunteer Spaces in Honduras." *Sociological Forum* 33(3): 690–711.

Schwarz, K. C. 2018. "Volunteer Tourism and the Intratourist Gaze." *Tourism Recreation Research* 43(2): 186–196.

Simpson, K. 2004. "'Doing Development': The Gap Year, Volunteer-Tourists and a Popular Practice of Development." *Journal of International Development* 16(5): 681–692.

Sin, H. L. 2009. "Volunteer Tourism—'Involve Me and I Will Learn?'" *Annals of Tourism Research* 36(3): 480–501.
Sin, H. L., T. Oakes, and M. Mostafanezhad. 2015. "Traveling for a Cause: Critical Examinations of Volunteer Tourism and Social Justice." *Tourist Studies* 15(2): 119–131.
Smith, V.L. 1978. *Hosts and Guests: The Anthropology of Tourism*. Philadelphia: University of Pennsylvania Press.
Su, C.P, and T.C Wu. 2020. "The Dark Side of Solo Female Travel: Negative Encounters with Male Strangers." *Leisure Sciences* 42(3–4): 375–392.
Unger, O., G. Fuchs, and N. Uriely. "Beyond the 'Tourist Environmental Bubble': Encounters with Locals and Destination Experiences of Business Travelers." *Journal of Travel Research* 59(8): 1493–1505.
van Eck, D., N. van Amsterdam, and M. van den Brink. 2021. "Unsanitized Writing Practices: Attending to Affect and Embodiment Throughout the Research Process." *Gender, Work and Organization.*
Vodopivec, B., and R. Jaffe. 2011. "Save the World in a Week: Volunteer Tourism, Development and Difference." *The European Journal of Development Research* 23(1): 111–128.
Wearing, S. L. 2001. *Volunteer Tourism: Seeking Experiences That Make a Difference*. Oxon: CABI.
Wearing, S., and S. Grabowski. 2011. "Volunteer Tourism and Intercultural Exchange." In *Volunteer Tourism: Theoretical Practical Application*, 193–210.
Wearing, S., M. Mostafanezhad, N. Nguyen, T. H. T. Nguyen, and M. McDonald. 2018a. "'Poor Children on Tinder' and Their Barbie Saviours: Towards a Feminist Political Economy of Volunteer Tourism." *Leisure Studies* 37(5): 500–514.
Wearing, S., J. Small, and C. Foley. 2018b. "Gender and the Body in Leisure and Tourism." In *The Palgrave Handbook of Feminism and Sport, Leisure and Physical Education*, 95–109. London: Palgrave Macmillan.
Zavitz, K. J., and D. Butz, D. 2011. "Not That Alternative: Short-term Volunteer Tourism at an Organic Farming Project in Costa Rica." *ACME: An International Journal for Critical Geographies* 10(3): 412–441.

NOTES

1. Sarah and Luz are self-selected pseudonyms used to protect participants' confidentiality; however, specific details of their identities and stories may identify them to other team members. They were made aware of this risk when agreeing to participate in this study.

2. The "tourist bubble" is a term Smith (1978), and many authors since (e.g., Unger et al. 2020; Adiyia et al. 2015; Carrier and Macleod 2005; Jacobsen 2003; Judd 1999), have used to describe tourists' tendency to stay amongst themselves, being physically in a foreign place but socially outside its culture.

3. *Purposeful Travel* is an NGO that facilitates trips in the Global South for those interested in ethical volunteering while traveling. Since its founding in 2006 the

organization has facilitated nearly 250 trips in 19 countries around the world. *Purposeful Travel* is a pseudynom used to protect the identities of the research participant. The organization agreed to participate in the research project and the result of the study were shared with them.

4. All names of local partners have been changed to protect confidentiality.

Chapter 3

"Skanky Stories"

Breaking Boundaries of Sexual Taboo in Women's Travel Narratives

Emily Falconer

INTRODUCTION

This chapter[1] focuses on reflective "travel stories" describing predominantly heterosexual sexual encounters during women's travel experiences. In short, this chapter argues that women engage with and subsequently re-tell stories of taboo sexual encounters as part of a wider discourse in performing particular identities through independent, global travel and adventure. The origins of this argument are based on a wider project from my doctoral research, *Embodiment and Emotion in the Experiences of Independent Women Tourists (2009–2012)*, yet this chapter will focus particularly on the intersection between collective, performative travel narratives, sexuality, gender, race, and class.

Contextualizing this argument within a broader field of tourist studies, this wider research explored the embodied and emotional experiences and reflections of women who engage in independent global travel. This form of tourism, which incorporates long term exploration of global destinations, low budget accommodation, "local" transport and a flexible itinerary, has been characterized as "backpacking tourism" in academic tourist studies (Richards and Wilson 2004; Hannam and Ateljevic 2008). Backpacking travel originates from earlier forms of "drifter," alternative tourism (Cohen 2004) where individuals traveled extensively through a variety of destinations, carrying their sparse luggage in backpacks. Despite its recent mainstream popularity, backpacking is still widely perceived as distinct from the mass tourist industry. As a rite of passage, it is argued that the backpacker uses "this limited period for unlimited freedom and hedonistic quest for intense, heightened states of exaltation and experience" (Cohen 2004: 51). There has been a significant rise in solo women engaging in this form of tourism (Wilson 2004). Gendered approaches to backpacking research have intended to address the gender-blind nature of previous literature, yet are similarly centered on women's self-empowerment and "journeys of independence" (Obenour 2005). Myers and Hannam (2008) refer to the association between women's growth in independence and their "empowerment," highlighting the benefits of travel to women's identities as free and liberated individuals.

Meanwhile, postcolonial critiques of backpacking travel associated with western privilege, whiteness and a fetishism for the "Other" has been long established (Teo and Leong, 2006; Lozanski 2010). There has, however, been limited research that draws together knowledge from tourist and backpacking studies with feminist theories on the (material) body, the politics of emotions, and the changing role of sexuality and feminist identities. Importantly, a further key characteristic of backpacking travel is the desire for close, embodied encounters with "dirt" (hot sweaty "local" buses, filthy toilets, grubby accommodation, cheap street food; Edensor and Falconer 2012). This quest for sensual immersion is linked to a perceived authenticity, and tolerating the hardship of "rough" travel is crucial to maintaining a cosmopolitan identity (Moltz 2006). This latter point is central to this chapter, as it is this "skanky" travel environment that feeds into women's "vulgar" narratives of filth, sexual taboo and crossing boundaries, as well as reproducing the racial stereotypes inherent in sex tourism.

Specifically, the focus of this chapter explores how women negotiate—and break through—embodied and sexual shame and taboo as a key part of their temporal process of learning to become an experienced traveler, creating particular forms of distinction (from both mass tourists and other women who do not engage in such sexual practices) through narrating their experiences. Transient sexualities within traveling spaces have been explored where we have seen that sexual behavior during travel is both often perceived as radically different from everyday life (Berdychevsky et al. 2010) whilst also proving (somewhat disappointingly) not as different to casual sex in non-traveling spaces as perceived (Frohlick et al. 2016). Either way, the expectations of sexual adventure in women's travel stories display a heightened sexuality that significantly reflects their journeys of exploration, experience, stimulus, whilst providing some entertaining "fodder." The narratives of women travelers as they re-tell stories of sexual behavior throughout their travels reflect the desires for extreme experiences to enhance their stories. This analysis will largely consider the "travel story" performed in focus groups, where women are able to perform a parody of "vulgar" sexuality as, by already being able to perform "tasteful" feminine sexuality (Skeggs 2001), they cross the boundaries into what could be perceived as taboo language. This performance enhances their perceived roles as temporary sexually deviant within traveling spaces.

CAPTURING THE "TRAVEL STORY": METHODS

The critical turn in tourist studies has had a profound impact on research methods, and has opened doors to new and innovative modes of critical tourist inquiry (Ateljevic, Pritchard, and Morgan 2007). To establish a more comprehensive understanding of the emotional, affectual and embodied experiences of women travelers required a methodology that both captured the performance of and worked alongside the narrative of the travel story. Four research methods were used to acquire the data for this thesis: qualitative interviews in the field (37 participants), participant observation in the field, four focus groups (16 participants) carried out in the UK with women who had returned from a recent period of travel and sensual diaries. These data draw upon two periods of fieldwork in popular backpacking enclaves: three months of travel in India in 2009 and three months of travel in India, Thailand, Malaysia, Australia, and New Zealand in 2010. Recruiting participants for the research required me to question my rationale with regard to what constituted suitable participants, and to what extent "traveling women" are a homogenous group (Mason 2002).

Identifying exact criteria for what constitutes "backpacking" or "independent" travel is problematic. However, as a starting point, participants recruited in the field were required to be traveling alone and independently (not as part of an organized tour, package or organization) for a minimum of three months. This was the only set criteria for participation in the study, allowing for any remaining diversity of women to emerge organically. This was a deliberate rationale, in order to accurately reflect the predominant demographics of those women who engage in this from of tourism, and who are willing to particulate in this research. As such, my sample demonstrated that those who were most likely to engage in this form of travel were predominantly of white ethnicity (89 percent), heterosexual, and educated to at least university degree level (95 percent held a university degree). Seventy-seven percent were European, 13 percent North American, 5 percent Antipodeans (Australia and New Zealand). Forty-eight percent were between 24 and 30 years of age, with 18–24 year olds as the second highest age group (47 percent). Only 2 percent were over 40 years of age, and just 5 percent held no University degree. Most participants were relatively wealthy, able to live for many months—and even years—without any income from employment, commonly using accumulated savings or inheritance. They did not have caring responsibilities or any dependents. That this sample personified a high degree of socio-economic, white and classed privilege is relevant to the overall argument of this chapter; that such privileged women have a greater degree of freedom to break boundaries of sexual taboo in both travel space and reflective narrative. Furthermore, the term "travel space" is used throughout this work to signify the symbolic locales in which these privileged forms of mobility take place. Therefore "travel space" is not always destination specific, but those spaces "in between"—such as backpacking hostels, transport, airports, and other transitional spaces.

This chapter will, for the most part, engage with data derived from the focus groups with women who were recruited in the UK who were collectively invited to reflect upon a period of independent travel, as this is where the humorous "travel sex story" was most apparent. Storytelling is central to both exploring personal and intimate practices in everyday life (consider Plummer's Telling Sexual Stories 2002) and—most crucially—in relaying to others exciting and entertaining tales from incidents "on the road" during backpacking travel. Plummer notes that telling sexual stories is part of a "wider discourse and ideologies abroad in societies, and they have much in common with all manner of other stories with differing foci—detective stories, travel stories, life stories, near death stories" (Plummer 2002: 6). Simultaneously, reflective narratives and storytelling has been researched in tourist studies as an integral part of how backpackers perform and make sense of their experiences (Noy 2004). The "Travel Story" becomes a scripted narrative, both through casual face-to-face interactions with friends, family and other travelers ("One time when I was on a night train in India . . . "), and increasingly more publicly through online travel blogs, social media and digital forums (Lean 2012; Moltz 2012). As such, when exploring the academic significance of backpacking travel, travel stories make up the significant body of narrative data for research into the experiences of participants. This is especially the case when drawing on reflective narrative methods, such as qualitative interviews and, moreover, focus groups. In contrast, data derived from participant observation and sensual diaries revealed more complex, and often difficult emotions and experiences of sexual (and other) experiences, capturing affective moments as they occur, rather than rehearsed, retrospective storytelling where participants maintain a degree of control over their narrative performance. Individual interviews, while more nuanced at times, often portrayed large sections of practiced, sexual travel stories as the participants wished to tell them. Yet the sexual travel story was nowhere more apparent than in the four focus groups, where sexual stories became increasingly intensified as the women interacted with each other. The retrospective and at times competitive travel story is a safe and legitimate space in which boundaries of sexual taboo are encouraged be broken.

Focus group methods have long been argued to address feminist ethical concerns about power and the imposition of meaning in feminist research (Wilkinson 1998) and have more recently been fore-fronted as suitable for "opening up" intimate discussions when researching sex and sexualities (Frith 2000; Berger and Lorenz 2016). Frith argues that focus groups not only reveal how participants engage with sexual activity, but how these activities are understood and made sense of within similar social groups (language used, emotional meanings). Recent attention to focus groups have shown suitability for articulating sensitive, taboo and potentially shameful topics such as intimate and embodied practices, dirt and "dirty topics" (Bowne 2016). Browne claims that the shared space of the focus groups allows such stories to flow, often cushioned by laughter, humor and "gossip" (2016: 199). The role of humor in the focus is therefore crucial to articulating the sexual travel story, and worked well as a collective exercise. Engaging with "dirty" topics both strengthens the role of the cosmopolitan embodied traveler and is central to the performance of travel narratives through the entertaining travel story, and the focus group provides a space where an amalgamation of sexual and travel stories can occur in particular ways. This is not to say that that the focus group method was free from problematic ethical considerations, and I will reflect on the potentially exclusionary and competitive nature of these group methodologies in the concluding discussion.

SEX AND TRAVEL

While research into "sex tourism" has tended to focus on male tourists who engage in commercial prostitution in tourist destinations, there has been a growing body of work that seeks to problematize this preoccupation (Ryan and Hall 2001). While "western women traveling unaccompanied by men have historically been linked to non-normative sexualities" (Frohlick, Gragojlovic, and Piscitelli 2016: 235), women engaging in sexual activity during periods of travel have become a growing area of research for tourism research. The key paradigms of this focus have shifted from an analysis of "female sex tourism" from the model of gendered, racial, and economic power dynamics, to the complex interplay of the role of casual sex in women's backpacking experience and the affective and embodied influences of spaces of travel on women's sensual and intimate relations. The umbrella term "sex tourism" has come to be understood to describe other forms of sexual relations in a tourist context which do not necessarily adhere to the politics of commercial exchange. Part of this research follows the highly contested debate about whether "female sex tourism" can fit into postcolonial, exploitative models of male sex tourism (Sanchez Taylor 2000, 2001, 2006; Pruitt and La Font 1995; Jeffreys 2003; Jacobs 2009, 2010). Sanchez Taylor (2006) argues that a double standard is applied by feminist academics to male and female tourist's sexual behavior. These claims assert that economically powerful traveling women engage with new gender roles and identities, but by doing so often reproduce the patriarchal structure of tourism (Pruitt and La Font 1995) by enjoying their dominant positions. More recently, Sanders-Mcdonagh (2010, 2016) has carried out research into female tourists in Thailand and Holland who have increasingly become consumers of the large-scale sex industry. There is an additional body of literature that focuses specifically on the complexities of women's sexual agency in a tourist setting. These more recent studies explore the importance of casual sex during the backpacking experience and how spaces of travel both facilitate and/or inhibit women's sexual behavior (Berdychevsky et al. 2010). Constructing themselves as "liberated, cosmopolitan feminist travelers" (Frohlick 2008: 131), it has been argued that women who engage in sexual activity in tourist spaces often equate the freedom of touristic mobilities with an intensification of their sense of sexual agency and empowerment. Traveling mobility provides a space for alternative sexual behaviors that require debates on female sexuality to be expanded into these different temporal and spatial dimensions. Research into women's sexual encounters during the traveling experience requires significant theoretical contributions from debates on the changing role of female sexuality and sexual liberation in feminist politics, to enquire about the extent to which women's sexual behavior during independent travel mirrors or contrasts with these changes and concerns. Theorizing the sexualization of culture in a postmodern era, Attwood notes that sex and sexuality reflect the broader conditions of our social world, that is: the injunction to be authentic, spontaneous, involved, hedonistic, a sensation-seeker, and yet to maintain control of our sexual selves; to self-fashion, remain detached and forever open to offers (2006: 89).

The above characteristics of sexual self bears resemblance to the existing literature on backpacking tourism, where we have seen that "authentic," "hedonistic," and pleasurable sensations are key motivations for this form of travel. It is necessary therefore to critique theories of the changing characteristics of sexuality within current trends in tourism, specifically with regard to backpacking tourism, from a feminist perspective. Carr and Poria (2010) put forward the necessity to widen the study of sex and the sexual in tourism and leisure studies in order to more accurately examine the wealth of sexual experiences, yet whilst they connect the "moral straitjacket" to origins of Catholicism (2010: 6) they do not make full use of feminist debates surrounding sexuality, or provide enough insight into the restrictive frameworks in which this "straitjacket" of morality operates, who is resisting it, or how. Carr and Poria (2010) attempt to provide a refreshed analysis of the sex and the sexual in tourism that move beyond the boundaries of "conventional morality," allowing research into leisure and tourism to emerge that transcends the shameful histories and taboos associated with sexual pleasure. The "agency pendulum" (Gill 2007) continues to provoke tensions within feminist debate, yet replacing women's sexual shame with discourses of pleasure and resistance remains a central to feminist debate. Most recently, the work of Frohlick, Dragojlovic, and Piscitelli (2016) illustrate a highly nuanced picture of foreign travel and transnational sex, arguing for greater complexity of insight into how these practices challenge heteronormativity. It is argued that western women as sexual subjects especially distort the boundaries of heteronormativity by exhibiting "unruly and nonconforming" behaviors (2016: 327).

BREAKING THE BOUNDARIES OF SEXUAL TABOO

Many reflective stories of the body incorporated overt sexual encounters during women's travel experiences. Women's impetus for independent travel spans across multiple motivations, from self-discovery to filling in a time in their lives between jobs, relationships and homes, and it would be wrong to suggest that the vast majority engage in casual sex with multiple partners. For those who do, however, the stories in my research highlighted sexual agency as a key element of their personal journey:

> Kelly (INT): [describing her sexual satisfaction with a young Indian man in Goa, India] We are both in our primes as someone was saying the other day. A women's prime is about 38 and for a bloke about 21! So fantastic! I mean you don't need to know the details but yeah! . . .
>
> Exactly. If you are both adults and no one is getting hurt. Consenting adults. Fuck it. (Kelly, 38, White British, Interview in field)
>
> Roxy (FG): [discussing how sexual promiscuity during travel results in different levels of sexual satisfaction, yet the primary goal is carefree fun]: Because we were drinking all the time, and you are meeting new people all the time, and so it's kind of like the excitement of "Oh you're new! Ooh actually you're great! Oh let's shag let's shag. Get in the hut! Get in the hut!" Or even, "Let's not bother getting in the hut!" [Roxy, 28, White British, Focus group in UK]

Here we see the "up for it" sexual expressions that Evans et al admit "at first glance seem to fulfill much of the ambition of second-wave feminism for an acceptance and celebration of active and engaged female sexuality" (2010: 115) but which also have provoked such powerful feminist concerns (McRobbie 2009). Levy outlines the powerful, unrepentant aggressive masculinity that comes with women who display such overt sexual power in her representation of the "female chauvinist pig": "She is post-feminist. She is funny. She gets it. She doesn't mind cartoonish stereotypes of female sexuality, and she doesn't mind a cartoonishly macho response to them." (2006: 93). Yet the performances of women travelers move beyond the concerns of over-sexualized "raunch" culture. The "empowered" sexual assertion in women's travel narratives show a greater reflective critique of the feminist debates than the "duped" women in the highly critiqued work of Levy and McRobbie, as the following discussions demonstrate.

Sylvia's interview is predominantly story after story of sexual encounters whilst traveling. She explicitly talks of the extreme excitement of "fucking" in a Catholic Croatian guest—house under paintings of the Virgin Mary, with the owners in the next room. She also tells me a story of self-pleasure:

> Sylvia (INT): Oh and one more thing because I think this is a good story and I don't think many people will be as honest with you, and it's repulsive! But not about shit and blood, it's about sex. I took real pleasure in working out weird places that I could masturbate. I only did it a few times, but it was like I just masturbated on an overnight train, or on the toilet of an airplane, I joined the mile high club by myself! It was part of the whole independence thing. Or like being on a cruise ship, I stowed away on a cruise ship in Norway at a really reduced fee, so I didn't have a cabin, only the floor, and I took high entertainment at sitting in the spa pool on the back of the cruise ship with all these really rich families and sitting on the spa jet and enjoying it. So it was like I am going to do something really naughty and self-contained . . . even more so then sleeping with other people . . . stuff I wouldn't do at home . . . So there you go! Some fodder! I can't believe some of the things I am telling you! (Sylvia, 26, White Australian, Interview in field)

Attwood notes that complaints on the objectification of women have "particular vehemence against images which draw on themes of women's sexual power, for example, those which can be read as expressive of lesbian desire, self-pleasuring or sexual power over men" (2005: 398). Sylvia's story displays a heightened sexuality that radically differs from the feminist concerns that female sexualization is geared toward the fantasy male gaze, and is further integrated into her traveling journey of exploration, experience, stimulus, whilst providing some entertaining "fodder." What is of further note, is how these pleasure-seeking discourses tie into adventure narratives so prominent in backpacking literature. The narratives of women travelers as they re-tell stories of sexual behavior throughout their travels reflect the desires for extreme experiences to enhance their stories.

The following extract indicates that Alison enjoys re-telling her stories of the uncharacteristic behavior that traveling allows, but acknowledges that this "madness" is a temporary role:

> Alison (FG): It's the wild, novel adventure of it all. Doing things you would never normally do. Like we got on a boat with guys who were coked up to their eyeballs and had mad sex I don't even remember! I mean it's all a bit much when I think back, and thank goodness I'm now home with my boyfriend—he's a doctor and so nice. But I do love talking about these sort of adventures with other girls, we just howl with laughter at the madness of it all. (Alison, 27, White British, Focus group in UK)

The association between outrageous sexual behavior with "madness" as Alison describes, relates to Walsh's study of women's transient hyper—heterosexuality, where "some laughed as they told me they 'went a bit mad' in reference to a higher number or frequency of sexual partners in Dubai" (2007: 524). Walsh notes that discourses of mental ill health are often used to describe women's approach to intimacy, where women often struggled between the discourses of healthy and unhealthy emotional (de)attachment in the context of feminine emotional norms in a culture of romance. Attwood has elsewhere celebrated the confident, self-possessed sexuality that challenge the dangers of associated madness, claiming "that sexual pleasure no longer indicates female evil or madness as it did in the medieval and moral discourses of the past is clearly a sign of progress" (2005: 397). Alison, however, appears to be using the concept of "madness" to boost her story telling credentials, adhering to the norms and expectations of backpacking theories and motivation where non-routine, hedonistic and "crazy" experiences sought after in order to stimulate the body and mind. When spaces of independent travel come together with the post-feminist articulation of sexual power and autonomy, the deep-rooted influences of both these discourses emerge in women's perceptions in a particular way. From the shelter of the focus group, where Alison has returned home to an environment of juxtaposed stability and her professional stable partner, she is now safely removed from her periods of "madness" to reflect upon discourses of mental health as a source of great humor. This contrasts with the tensions reflected on Dragojlovic's work into women travelers who had to negotiate "playing family" after having children with Balinese men in Indonesia as part of a traveling romance (2016). Where the mothers in Dragojlovic's work compete between discourses of heterosexual motherhood and nonnormative sexualities apparent in transient sex, no such tension is necessary for Alison who can safely return to "suitable" relationships and normative depictions of motherhood so starkly separate from her "mad times." To enjoy sharing stories of "madness" with others suggests a solidarity amongst women who have returned from the adult playground-like spaces of travel, where their anxieties surrounding their emotions, behaviors, cultural norms and tensions between the paradoxes of sexuality can be safely attributed to a period of "madness" that is widely acceptable, even desired, within the confines of travel spaces. Those who return home without a tale of profound madness will lack the relevant stories. Two points emerge here, firstly that they are open about their sexual pleasures and "crazy" behavior, and secondly are able to transcend the shame often associated with this kind of sexual behavior through the shared language and acceptance of the focus group. The language and sexual vocabulary of the women as they discuss sexual identity whilst traveling plays with the boundaries of both gendered and classed respectability in a conscious manner. In the focus groups, the women were aware of their changing roles, and able to perform a parody of "vulgar" sexuality as, by already being able to perform "tasteful" feminine sexuality, crossing the boundaries into what could be perceived as unacceptable language enhances their perceived roles as temporary sexually deviant within traveling spaces. Consider this narrative performance from Meena:

> Meena (FG): I have always been a bit of a whore-bag! [This provokes a roar of laughter from the focus group, who continue to laugh extensively.] But no I would probably say even more. . . . but I think it's . . . yeah it's great! I think it's easier when you're travelling because everyone's up for it generally. And you don't give a shit really. (Meena, 30, Mixed Race British, Focus Group in UK)

The above quote from Meena took place in one of the focus groups, where sexual stories and behaviors from traveling dominated the discussion. As discussed in the methodology section, the focus groups provided a forum in which women could tell a story of their sexual encounters, and reflect on that story within the group. What was prominent in Meena's particular focus group was laughter and camaraderie that accompanied the story telling. The laughter to Meena's self-description as a "whore-bag" indicates a humorous resistance to what could be perceived as a derogatory and offensive term to describe sex workers, women's sexuality and sexual practices. The boundaries of sexual vocabulary are crossed and accepted, first by Meena herself and then confirmed by the laughter from the group. This problematizes Attwood's figure of the "slut" as a sign of low social class and vulgarity among women who have been taught the importance of the "classy" sexually liberated female from popular culture. It can be argued that the term "whore-bag" fits into similar language used to traditionally describe women's deviant and classed sexualities, such as "slag and slut." Meena is aware of the classed nature of her term, and her breach of legitimate sexual femininity is a conscious act, and one designed to provoke similar reactions of shock, taboo and humor. Similarly, Meena's representation of the "whore-bag" questions Kitzinger's findings into white, young women in Glasgow (1995) who, she claims, walk the tightrope between sexually desirable femininity and "sluttish" reputations, where power inequalities are largely absent from their discourses.

Challenging this paradox, Meena reclaims the slang term "whore-bag" from the realm of unacceptable, vulgar feminine sexuality, displaying resistance to what Walsh describes as the "persistent gendering of morally loaded terminology" (2007: 523). Instead, she reclaims such uninhibited sexual agency through the use of this term, which she is confident will provoke entertainment within the specific space of the focus group, whilst simultaneously distancing herself from this. Frohlick (2008) refers to the tightrope between the "loose whore" and the sexually emancipated women and the tensions in female sex tourism, yet it would appear Meena is not unfamiliar with these dualisms. It is unknown whether Meena would describe herself as a "whore-bag" in all social spaces, in fear of evoking the wrong response. Yet it is evident that crossing boundaries of acceptable feminine sexual behavior and language is encouraged within these spaces, with these groups of women, who choose to claim the space of the focus groups to reflect on their sexual encounters in traveling spaces from the direction of empowerment and agency.

Raquel, Meena, and Roxy continue sharing sexual anecdotes throughout the focus group. They relate stories about other travelling women who they met, who had abandoned unsatisfactory sexual experiences during intercourse, deciding they had "had enough," or recalling how easy it is to "copoff" with someone in a beach bar. Each story is met with exclamations of affirmation, such as "brilliant!" and "that's hilarious!" confirming the humorous potential of such stories. Often, they would introduce an anecdote with wording such us "My favorite sex story is . . . ," and after one particularly explicit sexual tale Meena jokes: "I think that's a brilliant story, that's one for the grandkids!"

Aware of the classed and gendered disrepute attached to the sexually explicit nature of their behaviors and stories, these women play with ideas of respectable femininity and attribute traveling spaces to those where the "shameful" nature of female sexuality becomes a game of bravado. The fun derives not just from sexual encounters, but the re-telling of these stories, where pushing the boundaries of sexual taboos becomes in itself the entertainment. It is clear that the focus groups provide a space where great fun can be had in breaking these boundaries, and where the performance of the "vulgar whore" legitimizes their identities as women who have a high degree of power and control over their own bodies, pleasures and sexual freedoms.

That these tales are "one for the grandkids" plays on the traditional idea of the famous travel and adventure story which will pass through the generations, by highlighting the irony that women's "shameful" sexual histories will most likely never be documented through such stories.

The focus group provides the space in which to perform controlled and empowered sexualities in a world that still limits the possibilities of female sexual freedoms. Yet the classed (and often raced) status of women travelers gives them privileged access to such freedoms: to cross the boundaries of subversion without repercussions. Phoebe Waller-Bridge's BBC Three series *Fleabag* (2016–2019), which follows a nameless upper-middle class, English white woman's complex family relationships and lewd hyper-sexuality through dark comedy, has been deemed as the epitome of the postfeminist masquerade (Gill in Matos 2017) and "takes pleasure" in powerfully crossing the affective boundaries of discomfort, disgust and rebellious abjection for its viewers (Woods 2019). Others, through media reviews (Jones 2019), have highlighted the deeply classed dynamic and privilege inherent in playing with these depictions of obscenity, and question whether those without such a high degree of raced and classed privilege could play equally. Comparing *Fleabag* to *Chewing Gum*, which follows a working class Black British woman, Woods (2019: 7) refers to the work of Wanzo (2016) suggesting "the black precarious-girl cannot revel in debasement, abjection and 'grossness' like her white peer. This is because 'abjection haunts black bodies as a representational shadow' (2016: 45) through racist stereotypes of the black grotesque produced through cultural histories built from white supremacy." It can be argued that the majority of woman travelers are aware that a temporary dip into "grossness" will not stick long term to their bodies or identities. As Bev Skegg's notes, "femininity is fixed so other can travel" (2001: 300). Drawing back to the work of Desforges (2000) who notes that the independent and self-sufficient characteristics of independent travel are incompatible with femininity, Walsh (2007) argues that exaggerated performances of transient heterosexuality are still highly gendered within tourist spaces, and women's overt sexual behavior and language is received differently. Parodying what is perceived as male sex tourist behavior further became a subject assumed to cause shock and humorous delight within the focus groups. Roxy recalls a sexual relationship with a Thai man in Bangkok during her travels. Prohibited from entering her guesthouse due to the owner's disapproval of sexual relationships between tourists and Thais, Roxy's lover would hire a room in the "sex hotel," where guests pay an hourly rate. Roxy describes the experience to the group:

> Roxy (FG): So we had to go and rent room for an hour! It was hilarious! Quite embarrassing actually, because it was mostly older men and young Thai girls and then us! [Lots of laughter]
>
> Raquel: "Oh that's funny!"] And I remember there was this thing in the room, this weird sex thing, like a sort of contraption. . . . but anyway so we just like . . . and then he had a motorbike and we would scoot about and go and visit his family, and we just had such a laugh [Roxy, 28, White British, Focus group in UK]

Roxy's distinction between the other, male sex tourists in the hotel and "us" (herself and her Thai lover) is particularly significant. Although she enjoys parodying this behavior, it is clear that she remains distant from the image of the male sex tourist in Thailand. It is because of this assumed distinction, that mimicking the sexual conduct of other male tourists who use these hotels for sexual encounters with Thai women and girls becomes an "embarrassing" story, yet one that is confident will be met with empathetic humor from other traveling women who also maintain this distinction. The origin of the funny story lies in behaving like a notorious sex tourist, and gaining access to these illicit spaces in order to report back the "hilarious" environments and sexual apparatus other, male, sex tourists" use for their sexual pleasure. This access increases Roxy's exposure to the Thai sex industry, and enables her to gain insider knowledge of these private spaces in order to enhance her own experiences and subsequent travel stories. Roxy crosses the material and geographical boundary of the predominantly male space of the "sex hotel," yet refuses to embody the shame that accompanies the women's perceptions of male sex tourist behavior. Apart from the brief embarrassment that comes from being associated with these other tourists, Roxy and her sexual partner "just had such a laugh."

Disassociation with the notion of the sex tourist is paramount in literature into female sex tourism (Sanders-Mcdonagh 2010, 2016), yet the complexities into the narratives of women who engage in what could be perceived as a form of sex tourism remain limited in these writings. Roxy's story belongs to the theme of the humor and camaraderie of mirroring shocking and shameful behavior, without embodying the actual shame.

CONCLUDING DISCUSSION: "SKANKY STORIES" AND BREAKING BOUNDARIES

As Martin (1992: 97) exclaims, "Every taboo on something shameful has the potential for rebellion written in it." Independent travel is one such space in which to experiment with these rebellions with regard to gendered and classed perceptions and practices of sexual behaviors. Spaces of travel and the women who move through them disturb the now widely critiqued feminist literature on raunch, sexualized and porno-chic culture (Levy 2006: Walters 2010), as these concerns do not fully capture the complexities and resistances of women's body images when merged with the cosmopolitan codes of the backpacking body. To be eligible for independent travel, the sexualized female body conforms to different representations of class, gender and cosmopolitan distinctions. Gill (2007) argues that for women in post-feminist cultures sexualized self-presentation has become normative, but care should still be taken not to become too sexualized as this is framed in negative sexual discourses. This argument has so far taken an unforeseen turn with regard to women travelers, many who downplay their sexualized appearance in favor of "skanky," unattractive appearances that seemingly liberate them from the pressures of post-feminist sexualization, whilst simultaneously engaging in overt, autonomous hypersexual practices. Relating to both her personal hygiene "on the road" and her sexual behavior, Roxy exclaims that travel allows her to be "skanky in every sense of the word!" This choice of word carries heavy associations of bodies that are dirty, slovenly, "nasty and slutty." Interestingly, the travel story provides a space where any mixed feelings relating to "bad" or dangerous sexual exploits, fears of sexual violence, sexually transmitted diseases, or difficult or unwanted experiences were either omitted or transformed into a different narrative: "Bad sex" becomes yet another anecdote for a hilarious story. The notion that the more "skanky" the better is key, as more seemingly reserved and "clean" women are met with condescending attitudes, mirroring the adventure discourses of backing tourism more broadly.

Indeed with regard to female sexuality and breaking the boundaries of sexual taboo, the post-feminist articulation of the pleasures of the body can be simultaneously seen as a form of hedonistic indulgence (Attwood 2005), which is starkly similar to the excessive indulgence identified in the "experience hunger" of backpackers (Cohen 2004; Welk 2004). On the surface, the uninhibited sexual behaviors of traveling women confirms Cohen's (2004: 51) observation that "utter hedonism and experimentation" has replaced the original political motivations of the traveling drifter. Yet Cohen's analysis of the shift in traveling behaviors fails to make links with the substantial literature on female sexuality, in which this unashamed and indulged hedonism and experimentation is presented as a key feature of post-feminist, raunch culture (Levy 2006). Neither does this shift consider the breaking of sexual boundaries as part of a different political project: one of resistance itself, where women challenge past notions of passive sexuality in traveling spaces and push the boundaries of public displays of femininity. These sexualities can be perceived as both highly powerful and deeply classed.

Feminist literature that has employed Bourdieu and the notion of "taste" in studies of femininity and popular culture (Skeggs 2001: McRobbie 2009), has aligned tasteful femininity with classed distinction. Maintaining forms of distinction are certainly prominent in the discourses of women travelers. However, my findings show that narratives that evoke disgust through "vulgar" language and subject matter are an essential part of a parody through which women distinguish themselves from those who have not acquired such cultural capital from travel as well as their more "tasteful"—and thus less experienced—traveling counterparts. By accumulating these markers of identity from independent travel, women can comfortably afford to play with notions of (dis)tasteful sexuality to enhance their stories. Furthermore, women further play with the concept of "madness," walking the tightrope between the sane and rational sexualized woman and indulging in "crazy" traveling experiences, where escaping from their perceived realities seems appropriate and encouraged. The enormity of the feminist debate into female sexuality continues to emerge in the narrative accounts, yet this influence is internalized, resisted, and played with in a variety of different ways when intersected with narratives of traveling identities. Boundaries of classed and gendered sexual identities become broken and blurred, as women travelers perform their travel stories in conjunction with the powerful privileges that come with being able to travel. It is of further importance to situate these narratives in a postcolonial context of white neoliberalism.

Although less about a "colonial nostalgia" as Jacobs (2010) would attribute to a western imaginary of "exotic" heritage, the women travelers in my research sought to obtain a sexual agency which affirms their own privileges as white, heterosexual, western women performing a particular notion of empowered sexual freedom. Jacobs has researched women's sexual encounters in luxury resorts of Sinai, Egypt where the construction of Bedouin paradise meets the seductive imaginary of the lavish Orient for western women. It is here, argues Jacobs, that the "gendered and racialized objectification of the (Arab) Bedouin man . . . can be experienced most directly through the body in a sexual/romantic encounter in the safety of tourist space" (2009: 51). Eroticized landscapes and the seductive qualities of tourist spaces feed into the romanticized "sexual ethnic encounter" (Jacobs 2010: 35), where women are able to perform a particular version of femininity that transports them to an imagined sense of pre-modernity (Jacobs 2010). These studies build upon theories of female sex tourism where the sexual relations between white women and Caribbean (Frohlick 2008) and Egyptian (Jacobs 2010) men are heavily gendered and racialized and work with racial stereotypes of black sexuality, but adds an important multi sensual dimension to these debates, where the spatial, embodied and seductive environment of the tourist space intersects with cultural expectations, influencing women's sexual behaviors from a number of directions.

This colonial imaginary of the Orient has pervaded in films such as David Lean's Lawrence of Arabia (1962) as Jacobs notes and, more recently, in Michael Patrick King's *Sex and the City 2* (2010). The latter, which received widespread criticism for depicting racist imaginaries of Arab culture and Orientalism, follows four affluent white women from New York City as they revel in the exotic luxury resorts of Abu Dhabi, getting waited on by humble, male Arab servants and exhibiting a sensuous sexual agency which is starkly contrasted with that of covered, and (seemingly) pitifully repressed, Arab Muslim women. Indeed *Sex and the City* (HBO 1998–2004) was made notorious for its positive representations of strong female friendship and post-feminist rhetoric of powerful women who "have sex like men" (Meyer 2014). Others have written about the series and subsequent films through a postcolonial critique of a western romantic ideology that reinforces and privileges whiteness, neoliberal values and commodified lifestyles (Meyer 2014: Wilkes 2015). Situating these white, neoliberal representations of feminist and sexual freedoms within the geographical backdrop of the Middle East reflects Jacobs (2009, 2010) key arguments that the postcolonial encounter is one where privileged women can play out their racialized fantasies of Orientalism (Said 1978). Interestingly, the focus groups in my research provided a space where, much like the now well-known scenes from *Sex and the City*, "the girls" can get together to bond over comedic stories of both good and bad sex in a manner that is enjoyable and affirming. There stories were of transient, "no-strings" sex which rebuffed the "romance tourism" attributed to notions of femininity (Pruitt and La Font 1995). As such, the focus group became the space where particular identities emerged and were subsequently reinforced: firstly that of the sexually confident, independent woman, and secondly, of the experienced backpacking traveler who was able to entertain her peers with wild and tantalizing stories. The narrative data from the focus groups somewhat contrasted that of the individual interviews, where a more balanced account of traveling experiences revealed a greater multiplicity of feeling, including vulnerability, discomfort and at times fear that emerged during in their traveling encounters and sexual relations. It is important to remain critical of the focus group space as inherently safe and supportive, despite the general agreement in feminist methodologies that group discussion is appropriate for talking about sex. Frith (2000: 284) provides a note of caution: "the dynamic and often intimate discussions which take place in focus groups can lead participants to reveal more information than they intended and to say things they later regret. Such over-disclosure may be particularly problematic if members of the group are not strangers but have a continued relationship outside of the group." The ethics of encouraging over-disclosure is particularly problematic when a space is specifically set up to share the best and "most juicy" travel story about breaking boundaries of sexual taboo. This can potentially foster a culture of competition and rivalry (and thus over-disclosure) between those who possess and are willing to perform "shocking" stories, and act as a means of exclusion and deep discomfort for those who do not. Be that as it may, for the most part—at least on the surface—the focus groups became the site of production for the cultural capital that is so deeply ingrained in traveling identities, as well as reproducing that familiar scene from countless film representations of white, neo-liberal women with much in common bonding over stories of heterosexual, hypersexuality in an affirming space.

However, the findings in this chapter remain limited with regard to the transgressing of gendered boundaries. Most recent urgings from Frohlick et al. (2016) support the need to be cautious in depicting women travelers as sexual transgressors, by focusing on the complexities of their own emotional and embodied accounts These narratives appear to liberate traveling women from the confines of femininity without the emotional burden of embodied and sexual shame, but only within the new confines of specific spaces, such as the retrospective interview or focus group, in which they have the power to disinvest from particular codes of femininity. Shame is not absent from these performances and narratives, but becomes embodied and re-negotiated as part of a complex emotional cycle. In this way, playing with the boundaries of femininity in traveling spaces is presented as a fun and joyful process of gendered resistance. What is central to this argument is the performative power of the reflexive travel story, where women are able to perform their desired gendered, traveling identities. Yet the non-reflexive, emotional embodied experiences (such as fears or experiences of sexual violence) which contradict these articulated performances do not as yet make it into these narrative accounts. Hopkins (2007) highlights that established critiques of the use of focus groups in human geography include only touching on shallow and limited insights, with certain perspectives being withheld from the group. Yet the providing such a space in the case of the sexual travel story shows how powerful these stories are in creating women's shared traveling identities and desire to break through boundaries.

Breaking the narrative boundaries of sexual taboo have relocated current feminist dilemmas of female sexualization to an alternative geographical space; indeed traveling space is one such arena in which a high degree of critical autonomy can be found with regard to breaking boundaries that go beyond the binaries of choices and constraints that emerge in feminist literature. However, these boundaries are not transcended without impediments, and are limited to the narrative performances that women select as part of their reflections and stories of entertainment. Gill argues that "practices that would have once attracted criticism from feminists are presented as playful, resistant or at least as the outcome of active, knowledgeable deliberation" (2007: 74). My analysis of the playful nature of women's travel narratives remains cautious of this argument, and situates the critical and resistant narratives of traveling women who break multiple boundaries in an alternative context, in order to address the dilemmas of agency and constraint in greater scrutiny.

More importantly, the "skanky stories" of this chapter reflect the postcolonial privilege of backpacking travel more broadly; it is this sensual and intimate immersion with the "Other" which marks the desire for closer, more authentic experiences (Lozanski 2010). As discussed, backpacking travel has been characterized as closely linked with embodied hardship, tolerance of discomfort and dirt. Notions of backpacking cosmopolitanism and a sense of global citizenship are embodied and performed through a body that is "flexible, adaptable and literally open to the world" (Molz 2006:7). "Skanky stories" of the body include not just those of sex, but accounts of illness, bodily fluids, adventurous street food, suffocating humidity, filthy toilets, dirty hard beds and uncomfortable transport that demonstrate how travelers embody a sense of tolerance as a mark of distinction from mass tourists who reside in luxury resorts (Edensor and Falconer 2012: 78). This is also apparent in literature on Slum tourism, where tourists desire a "dirty," sensual overload of visceral poverty in which they can feel a "thrilling sense of disgust" (Holst 2018). The crux of these stories connects the "filth" inherent in this particular form of travel with the "filthy" stories of sexual taboo. Multiple boundaries are crossed and broken, yet it is the women who embody high degrees of classed and raced privilege who are the freest to travel, geographically and symbolically, across these borders.

ACKNOWLEDGMENTS

I would like to thank all the participants who took part in this research and shared their intimate stories and reflections. I also extend my thanks to the anonymous reviewers and editors of *Gender, Place and Culture*, whose thoughtful feedback of earlier versions enabled me to revisit the data critically and provide a more robust analysis.

REFERENCES

Ateljevic, Irena, Pritchard, Annette and Morgan, Nigel. (2007) *The critical turn in tourism studies: Innovative research methodologies*. Oxford: Elsevier.

Attwood, Feona. (2005) "Fashion and passion: Marketing sex to women." *Sexualities*, 8(4), pp. 392–406.

Attwood, Feona. (2006) "Sexed up: Theorizing the sexualization of culture." *Sexualities*, 9(1), pp. 77–94.

Berdychevsky, Lisa, Poria, Yaniv and Uriely, Natan. (2010) "Casual sex and the backpacking experience: The case of Israeli women." In Carr, Neil and Poria, Yaniv (eds.), *Sex and the sexual during people's leisure and tourism experiences*. Newcastle upon Tyne: Cambridge Scholars, pp. 105–118.

Berger, Ronald and Lorenz, Laura S. (2016) "Talking about Sex: Focus Group Research with People with Disabilities." In *Disability and Qualitative Inquiry: Methods for Rethinking an Ableist World*. Oxen: Routledge.

Carr, Neil and Poria, Yaniv. (2010) *Sex and the sexual during people's leisure and tourism experiences*. Newcastle: Cambridge Scholars.

Cohen, Erik. (2004) "Backpacking: Diversity and Change." In Richards, Greg and Wilson, Julie (eds.), *The Global Nomad: Backpacker Travel in Theory and Practice*. UK: Channel View Publications, pp. 43–60.

Desforges, Luke. (2000) "Traveling the world: Identity and travel biography." *Annals of tourism research*, 27(4), pp. 926–945.

Dragojlovic, Ana. (2016). "Playing family": unruly relationality and transnational motherhood. *Gender, Place & Culture*, 23(2), pp. 243–256.

Edensor, Tim and Falconer, Emily. (2012) "Sensuous Geographies of Tourism." In Wilson, Julie (ed.), *The Routledge Handbook of Tourism Geographies*. London: Routledge: pp. 74–81.

Evans, Adienne, Riley, Sarah and Shankar, Avi. (2010) "Technologies of sexiness: Theorizing women's engagement in the sexualization of culture." *Feminism & Psychology*, 20(1), pp. 114–131.

Frith, Hannah. (2000). "Focusing on Sex: Using Focus Groups in Sex Research." *Sexualities*, 3(3), 275–297.

Frohlick, Susan. (2008) "'I'm more sexy here": Erotic subjectivities of female tourists in the 'sexual paradise" of the Costa Rican Caribbean." In Priya Uteng, Tanu and Cresswell, Tim (eds.), *Gendered mobilities*. Hampshire: Ashgate Aldershot, pp. 29–142.

Frohlick, Susan, Ana Dragojlovic, and Adriana Piscitelli. (2016) "Introduction: Foreign travel, transnational sex, and transformations of heterosexualities." *Gender, Place & Culture*, 23(2), pp. 235–242.

Gill, Rosalind. (2007) "Critical Respect: The Difficulties and Dilemmas of Agency and Choice for Feminism." *European Journal of Women's Studies*, 14(1), pp. 69–80.

Hannam, Kevin and Ateljevic, Irena. (2008) *Backpacker tourism: concepts and profiles*. Clevedon: Channel View.

Holst, Tore. (2018) *The Affective Negotiation of Slum Tourism*. London: Routledge.

Hopkins, Peter. (2007). "Thinking critically and creatively about focus groups." *Area*, 39(4), pp. 528–535.

Jacobs, Jessica. (2009) "Have sex will travel: Romantic 'sex tourism' and women negotiating modernity in the Sinai." *Gender, Place & Culture*, 16 (1), pp. 43–61.

Jacobs, Jessica. (2010) "Re-branding the Levant: contested heritage and colonial modernities in Amman and Damascus,." *Journal of Tourism and Cultural Change*, 8 (4), pp 316–326.

Jeffreys, Sheila. (2003) "Sex tourism: Do women do it too?" *Leisure Studies*, 22(3), pp. 223–238.

Jones, Ellen. (2019) "*Fleabag* is a work of undeniable genius. But it is for posh girls." *The Guardian*, April 20.

Lean, Garth. (2012) "Transformative travel: A mobilities perspective." *Tourist Studies*, 12(2), pp. 151–172.

Levy, Ariel. (2006) *Female chauvinist pigs: Women and the rise of raunch culture*. New York: Free Press.

Lozanski, Kristen. (2010) "Defining 'real India': Representations of authenticity in independent travel." *Social Identities*, 16 (6), pp. 741–762.

Martin, Emily. (1992) *The woman in the body: A cultural analysis of reproduction*. Boston: Beacon Press.

Matos, Carolina. (2017). "Rosalind Gill: 'We don't just want more cake, we want the whole bakery!'" *MATRIZes: Revista do Programa de Pós-Graduação em Comunicação da Universidade de São Paulo*, 11(2), pp. 137–160.

McRobbie, Angela. (2009) *The aftermath of feminism: Gender, culture and social change*. London: Sage.

Meyer, Michaela. (2014) "The fantasy of *Sex and the City*: A postcolonial critique of the romantic imaginary, consumerism and globalization." *Women's Studies*, 43 (4), pp 425–439.

Molz, Jenny. (2006) "Cosmopolitan bodies: Fit to travel and travelling to fit." *Body & Society*, 12(3), pp. 1–21.

Molz, Jenny. (2012) *Travel connections: Tourism, technology, and togetherness in a mobile world*. London: Routledge.

Myers, Linda and Hannam, Kevin. (2008) "Women as backpacker tourists: a feminist analysis of destination choice and social identities from the UK." In Hannam, Kevin and Ateljevic, Irena (eds.), *Backpacker tourism: Concepts and Profiles*. Clevedon: Channel View Publications, pp. 174–187.

Noy, Chaim. (2004) "This trip really changed me: Backpackers' narratives of self-change." *Annals of Tourism Research*, 31(1), pp. 78–102.

Obenour, William. (2005) "The Journeys of Independence for Female Backpackers." *Tourism Review International*, 9(2), pp. 213–227.

Plummer, Ken (2002). *Telling sexual stories: Power, change and social worlds*. Routledge.

Pritchard, Annette and Morgan, Nigel. (2000) "Privileging the male gaze: Gendered tourism landscapes." *Annals of Tourism Research*, 27(4), pp. 884–905.

Pruitt, Deborah and LaFont, Suzanne. (1995) "For love and money: Romance tourism in Jamaica." *Annals of Tourism Research*, 22(2), pp. 422–440.

Richards, Greg and Wilson, Julie. (2004) *The global nomad: Backpacker travel in theory and practice*. Clevedon: Channel View Books.

Ryan, Chris and Hall, Michael. (2001) *Sex tourism: Marginal people and liminalities*. London: Routledge.

Said, Edward. (1978) *Orientalism.* London: Penguin.
Sanchez Taylor, Jacqueline. (2000) "Tourism and 'embodied' commodities: Sex tourism in the Caribbean." In Clift, S. and Carter, S. (eds.), *Tourism and sex: Culture, commerce and coercion.* London: Pinter, pp. 41–53.
Sanchez Taylor, Jacqueline. (2001) "Dollars are a girl's best friend? Female tourists' sexual behavior in the Caribbean." *Sociology*, 35(3), pp. 749–764.
Sanchez Taylor, Jacqueline. (2006) "Female sex tourism: a contradiction in terms?" *Feminist Review*, 83(1), pp. 42–59.
Sanders, Erin. (2010) "Situating the female gaze: Understanding (sex) tourism practices in Thailand." In Hardy, Kate, Kingston, Sarah, and Sanders, Teela (eds), *New Sociologies of Sex Work.* Ashgate: Surrey.
Sanders-Mcdonagh, Erin. (2016) *Women and Sex Tourism Landscapes.* London: Routledge.
Skeggs, Beverley. (2001) "The toilet paper: Femininity, class and mis-recognition." *Women's Studies International Forum*, 24(3–4), pp. 295–307.
Teo, Peggy and Leong, Sandra. (2006) "A postcolonial analysis of backpacking." *Annals of Tourism Research*, 33(1), pp. 109–131.
Walsh, Katie. (2007) "'It got very debauched, very Dubai!' Heterosexual intimacy amongst single British expatriates." *Social & Cultural Geography*, 8(4), pp. 507–533.
Walter, Natasha. (2010) Living dolls: The return of sexism. London: Virago, 2007.
Welk, Peter. (2004) "The beaten track: Anti-tourism as an element of backpacker identity construction." In Richards, Greg and Wilson, Julie (eds.), *The global nomad: Backpacker travel in theory and practice.* Clevedon: Channel View Publications, pp. 77–91.
Wilkes, Karen. (2015) "Colluding with neo-liberalism: Post-feminist subjectivities, whiteness and expressions of entitlement." *Feminist Review*, 110(1), pp 18–33.
Wilkinson, Sue. (1998) "Focus groups in feminist research: power, interaction, and the co-construction of meaning." *Women's Studies International Forum*, 21(1), pp. 111–125.
Wilson, Erica. (2004) "A 'journey of her own'?: The impact of constraints on women's solo travel." PhD diss., Southern Cross University.
Woods, Faye. (2019) "Too close for comfort: Direct address and the affective pull of the confessional comic woman in *Chewing Gum* and *Fleabag*." *Communication, Culture and Critique*, 12(2).

NOTE

1. This chapter originally appeared in *Gender, Place & Culture*, 27(9). Permission has been acquired for its inclusion in this edited text.

SECTION II

Hosting

Chapter 4

Women's Work and Tourism in Negril, the Capital of Casual

A. Lynn Bolles

This chapter first looks at Jamaica's tourism overall and then winnows down to Negril, the "capital of casual" located on the western most tip of the country. I contend that tourism in Jamaica is a predominately "female" industry in terms of a product resulting from the expansion of neoliberalism that garnered capital private and public investments to the sector. The workforce in tourism is composed primarily of working-class women who rely on the sector for finances to support the welfare of their children. Consequently, the precariousness of the tourism industry, hindered by hurricanes, global economic down turns, as well as shifts in political largess, hover over the monetary opportunities for the women in Negril. Further, often the overdue material gains for women are overshadowed by gender systems that are inequitable in terms of the status and benefits.

The tourist sector is the second largest employer of women outside of categories of "own account" or self-employed. Many of those self-employed women can be found working in the tourist sector as road-side venders ("higglers" traditional agricultural producers and sellers) and sex workers who usually do not get counted in official documentation. But while social hierarchies and social differences in society surface in tourism, the Jamaica Tourist Board amends its promotions to deflect Jamaica's major problems such as crime and drug related violence by promotional tags, such as the early 1980s of "Come back to way things used to be" that harked back to a legacy of "colonial bliss" that is romanticized for tourist consumption. As a service sector, tourist work fits neatly into Jamaica's economic hierarchy. Tourism's gendered segmented labor market provides service work in establishments that pay taxes, such as hotels, banks, and restaurants, and that are counted and considered a service-producing tradable in the parlance of the GDP. Work in the "Elementary occupations," such as roadside vendors and sex workers, is often not counted, but critical to the tourist industry.

And then, there is the overall image women's tourist work as the chambermaid and the sex worker that solidifies the situation. In her classic text, *Bananas, Beaches and Bases*, Cynthia Enloe (1989, 34) remarked how in the 1970s, Caribbean nationalists complained that their government's pro-tourism policies had turned their society into a "nation of busboys." Sarcastically, Enloe suggested that perhaps 'the nation of chambermaids' did not have the same mobilizing ring in their ears." After all, a woman who has traded work as an unpaid agricultural worker for a hotel cleaner has not lost any of her femininity." This quote recalled two points. First is the concept of femininity aligned with the standard Eurocentric, dominant cultural notion of what was considered a gender norm. Second was the assumption that the labor of a working-class woman was deemed of such low value that no matter of the location, it was a gendered normative occupation. Nevertheless, the success of the tourist industry is categorically based on low waged female workers. With the explosion of the tourism in the Caribbean and particularly in Jamaica, the work of this "chambermaid" became the bedrock of this industry. Furthermore, if success in tourism is engineered by good service, hospitality, and professionalism then the role that women play was critical to that accomplishment.

RACE, CLASS, AND EDUCATION

Before going further, there are things to be said about the Jamaican social system that are based on the interconnections of skin color, class, and access to education. In the Jamaican context, skin color (phenotype) and class are historically connected, with people of darker hue (the majority) occupying the low strata of society. However, a proven key toward upward mobility in Jamaica continues to be education. Accordingly, following independence in 1962, the numbers of citizens with an elementary education increased dramatically when primary school up to grade six was made compulsory. Although technically free, secondary education remains costly for low-income households. It was not just the "school fee" but the required uniforms, school supplies, registration, and examination expenses that drove up the cost. In 2015, the bill for one child in secondary school amounted to US$300. That figure accounted for almost 20 percent of a Jamaica per capita income. Those are just some of the contributing factors responsible for the high rates of students dropping out of secondary education (wenr.wes.org). Although there was a high percentage (97 percent) participation rate in elementary schools, 2017 data showed that the percentage dropped to 60 percent enrolled in secondary school education (wenr.wes.org). Those who advanced their education through technical training and university levels illustrated how wide the gap can be. Plus, access to education continued to be heavily skewed toward higher income households in the Kingston Metropolitan Area (ibid.). In the 2010s, the Government of Jamaica's contribution to education rose from three to six percent of the GDP, and by 2017/18, increased an additional 3.9 percent. Caribbean Advanced Proficiency Exams (replacement of British "A" levels) and Technical and Vocational Training (T-VET) credits provided entry into skilled and managerial rank employment. An Associate of Arts in Tourism earned at a T-VET institution, such the community college in Montego Bay, became a valued credential. There are five levels of achievement in the T-VET program with the managerial and professional work at the apex. In 2017 there were 66,000 students enrolled at T-VET institutions (ibid.).

In addition to education, other obstacles still faced by women include skin color gradation that hark back to the days of slavery but with a contemporary twist. Women considered "brown" can be light skin colored with somatic features (light eye color, hair texture) or dark skinned. "Brown" as a social term also referred to middle class status. In 1938, Zora Neale Hurston ([1938] 1981, 76) made this observation, "If a woman is wealthy, of good family and mulatto, she can overcome some of her drawbacks. But if she is of no particular family, poor and black she is in a bad way indeed in that man's world." In contemporary times, the "color bar" can be lifted as new economic and educational opportunities become more available to a larger number of women than before. However, the options are fairly limited for most women. Since these women are poor or working class and black, the nexus of class/race nexus is further apparent.

The combination of class and color constraints are historically based but are held tight in Jamaica's highly stratified social system. A seminal work on social mobility in Jamaica by Derek Gordon (1987) was prophetic when he noted the following, "researchers must confront the paradox of large-scale social mobility generated by the opening up of new positions coexisting side by side with gross and perhaps, even widening inequalities of opportunity between the minority at the top and the majority at the bottom of the social order." Class and color do count in factoring mobility intersecting with gender. As a host of Caribbean scholars and their feminist colleagues from abroad argue, sexism constrains and limits women's access to opportunities in Jamaica. The on-going social and economic struggles faced by women in Jamaica also have a history of those who persevered despite these obstacles. An example is none other than the former Prime Minister of Jamaica the Honorable Portia Simpson-Miller (2006–7; 2012–16) and other outstanding women who are noteworthy in a variety of fields of endeavor throughout society.

TOURISM AND WOMEN WORKERS

Women Working in Tourism

Early tourism research on women focused on the hosts and new workers in development programs. Margaret Swain (1995) brought the importance of including women housed in a gendered frame as members of communities that are now tourist sites. For example, Mary Castelberg-Koulma (1991,197) noted, "Few studies address themselves to the effects that tourism has on the local female population." Or as Dallen Timothy (2001, 246) reminded, "Gender differences are obvious in the production of tourism in developing countries . . . and women's work resembles the types of work they have traditionally done in the domestic setting." Heather Gibson (2001, 28) reviewed a body of literature e.g., Kinnard, Kothari and Hall (1994) that focused on "gendered hosts," who are females confined to jobs such as chambermaids, laundresses, and waitresses usually characterized by low pay, low job security and powerless experienced in those tourist sites. Kinnard and Hall (1996, 96) argued that any gender analysis should address the differences in the quality and type of available, the differential access of women to employment opportunities, and seasonal fluctuations and the existing anew gendered divisions of labor. Ten years later, the focus on women as workers and as members of communities now tourist sites was exemplified by Kristen Ghodsee (2005) writing about Bulgaria, Denise Brennen's work in the Dominican Republic (2004), and Nandine Fernandez's (1999) study on race and tourism in Cuba. Those scholars not only examined the factors pointed out by Kinnard and Hall but put those traditional female jobs in a complex of opportunities afforded to women. They also shed light on how women instrumentalized those activities for their own advantage and understood the consequences of those actions.

Change comes slowly even when material conditions make it possible for women to have greater access to resources in society than ever before. Regardless of their class position, a woman's wage, combined with that of her male spouse/partner (who, more than likely will earn more than her), creates a dual-income household that often generates greater earnings for that domestic unit. But as Dunn and Dunn (2002) found, there was a distinctive disparity in income earned by men and women in Jamaica; males earned three times more than females in like-income. As of 2020, that wage gap increased so that in Jamaica for every US$100 a male earned, a woman earned US$60 (statista.com 2021). Often, there were very poor women who eked a very minimal existence. The overall condition of living in abject poverty was perhaps made more acute in the face of the tourists who ignored stretched-out hands as women begged from the side of the road or in front of supermarkets.

The UNWTO joined forces with UN Women and other UN agencies and organizations to compile data on the contribution that tourism was making toward the UN Sustainable Development Goal 5—Achieving Gender Equality and Empowering all Women and Girls. The second edition of this report looked at key factors that contribute to gender equality in the tourism sector that pinpoint challenges, and identify ways to lessen inequality, while capturing the potential that tourism offered to women participants. There were five thematic areas—employment, entrepreneurship, education and training, and leadership—all demonstrating how gender equity and women's empowerment in tourism would look (UNTWO 2019, 1). Case studies provided four key objectives for tourism industries—digital platforms and technology, hotels, and accommodations, tour operators, and community-based tourism. UNTWO reported that most of the global workforce in tourism is female; 54 percent women were in the sector while 39 percent were in the broader economy. The wage-gap was indicative of the low skill work usually performed by women, who earned 14.7 percent less than men. At the same time, tourism offered women greater opportunities for leadership roles. Twenty-three percent of tourism ministers were female, compared to 20.7 percent of all government Ministers.

The UNWTO report noted three major points for discussion:

1. Women were challenging gender stereotypes in all sectors and entering occupations usually dominated by men.
2. Technology was an important factor empowering women when coupled with training opportunities that encouraged women's entrepreneurial entry into tourist markets.
3. Policy-makers were becoming more aware of gender equity in tourism, putting in place measures that helped to ensure women shared in the benefits that tourism can provide.

The UNWTO report is an important referent for nations just coming on stream into the global community of tourism. However, while Jamaica is considered a mature tourist destination with a 125-year history of being in the business, there are critical sticking points that are mirrored in the UNWTO report. These issues include the continued opening of employment opportunities across all sub-sectors of tourism, the use of technology and the digital environment as an avenue for women's empowerment as entrepreneurs, and the long-term emphasis on gender equity not only in tourism but throughout society. At this juncture, all of these issues are clouded by the 2020/21

COVID-19 pandemic crisis. Secretary-General of the United Nations Antonio Guterres said, "In the midst of a global pandemic, one stark fact is clear: The COVID-19 crisis has a woman's face." Adding UNTWO Secretary-General Zurab Pololikashvili "Tourism is a proven driver of equality and opportunity. This unprecedented crisis has hit our sector's women fast and hard, which is why gender equality and empowerment must be centre stage as we work together to restart tourist and accelerate recovery" (WTO, 2021). Policy makers were encouraged to address women's economic and social insecurity in the face of the rise in unpaid care work and domestic violence (ibid.). The Government of Jamaica's Ministry of Labor (ilo.org) provided General Grants and Tourism Grants for small businesses, craft market vendors, and beauty therapists among others if they had applied for status through the Tourism Product Development Company. How effective those grants were for women who own or work in those enterprises will a topic for future study.

Higglers and Entrepreneurs

Like other Eurocentric models addressed by the UN, much of Jamaica's normative gender system is predicated on the dominate ideology of men as breadwinners and women as homemakers. One of the UN strategies for women's empowerment centered around women's entrepreneurship as an avenue toward gender equity. For Jamaica, entrepreneurship was encoded during the days of enslavement. Afro-Caribbean men, but mostly women grew provision grounds and traded those foodstuffs in the internal market (Mintz [1955] 1974, 180–213). Due to that same system, Black women, the descendants of the enslaved are seen as astute businesswomen. Writing about contemporary Barbados, Carla Freeman (2014, 1) asserts, "Entrepreneurialism . . . is not simply a mechanism of self-employment—a vehicle for income generation, an economic matter of business, that is entrepreneurship in a narrow sense—but a subtler, generalized way of being and a way of feeling in the world." As argued artfully by Carla Freeman in *Entrepreneurial Selves*, the role of women's entrepreneurship and small business is very much part of Barbadian society coming out of the seventeenth-century British plantation system. Women's entrepreneurship now morphed as an avenue for Black upward social mobility and part of the ever-evolving neoliberal capitalist investment venture.

In my study of Negril, tourism was an already established neoliberal development that framed the local economy. Therefore, the emphasis was not on the connection of middle-class respectability as an appendage of entrepreneurship nor on global neo-liberalism. Instead, the focus is on the history of higglers as micro-entrepreneurs and how this tradition is now used as a business model by poor and working-class women in Negril's tourism. Further, this long-standing tradition was utilized by Negril's women's entrepreneurs because it relied on access to consumers, a distribution system of goods and services, strategic use of time, and inventory management. Jamaican higglers are called just that, while in the Eastern Caribbean they are referred to as hucksters and traders.

Long-Standing Tradition

After the emancipation of the enslaved, the wider society considered higglers in a negative light. Collectively, higglers were fixed as being shrewish, loud, boorish, aggressive, and exhibiting behaviors associated with lower classes of society. In the Jamaican context, this assumed skin color too. Skin color was a critical factor as much of the working class and the poor are dark skinned. Part of the higgler identity also had to do with what was deemed socially acceptable deportment for women. But regardless of their business acumen, keen awareness of trends and financial soundness, higglers were not deemed proper. In recent years, feminist social scientists and activists have raised the status of the higgler in the eyes of many in the society. A case in point is Gina Ulysses's *Downtown Ladies* (2007) an award-winning ethnography of these informal importers and new age higglers. Nonetheless, contemporary small businesswomen who are not higglers are still tainted by the traditional image of market women, due to their aggressiveness, even though these modes of behavior are deemed smart and tough when attributed to a businessman. Businesswomen may engage in other kinds of commercial and business services that are highly technical and capital intensive, but they still remain in inequitable positions due to the general ideology regarding women in the economy (Carter and Cannon 1992). Of course, there are women who have been extremely successful in business, and Jamaica has quite a number of them too. Each of those women's rise to the top included not only surpassing basic obstacles, but also those that were gender related, including proving way beyond reasonable doubt that they were worthy, despite being a female. Again, this perception relates to the contradictory nature of inequality inherent in Jamaica's gender system.

Taking Care of Business

Evidence from both UNWTO reports and research from the University of the West Indies indicate the direction that governments, stakeholders, and women themselves need to take to successfully compete in today's tourist marketplace. These themes targeted the importance of technology, education and training, leadership, service, hospitality, and professionalism to ensure women's empowerment. The narration of the interlocutors featured in my research exemplified the extent that Negril is moving forward, and in some instances leading the way regarding the UNTWO themes. Given that Negril was late in coming into heavy neoliberal investment schemes, the time period of my two decades plus study is illustrative of tourism development as an ongoing process in Negril.

Introducing the Interlocutors

Negril was a fishing village whose inhabitants settled on this slip of marsh land following emancipation (1834–38). Tall coconut trees provided work for women who boiled and took the oil to local markets of Savannah La Mar. By the 1940s however, the coconut tree blight denuded the multiple stands of trees leaving open spaces and clear white sandy beaches. It took another almost leisurely paced thirty years for Negril to develop into a major tourist destination that features the seven miles of white sandy beach. My ethnographic research looks at how women's work factors into this evolution in Negril by focusing on women's work across a spectrum that in fact made Negril "the capital of casual."

One of the goals of my study was to elicit interviews and carry out conversations with women tourist workers across wage earning categories. Listed are six groupings that indicated the overall type of work and the number of women who did that work:

1. Entrepreneurs including traditional higglers but also an owner of a dive shop, mini-store owner, and hair braiders.
2. Cottage and Small Hotel owners.
3. Professionally trained accountants, clerks, and multilingual hospitality staff.
4. Housekeeping.
5. Food and Beverage cooks, chefs, table captains, food and beverage managers.
6. Entertainment activity directors, professional dancers, tour guides.
7. Craft Vendors both general and specialty craft artisans.

Indicative of the gamut of occupations, educational levels ranged from grade 3 to 6 in elementary, associate degrees to bachelor's degrees, an BFA and almost an MBA. The women's ages ranged from sixteen to seventy. For some workers, such as the hair braiders and cooks, their jobs were learned as girls in their natal homes as part of their socialization. The second-generation hotel owners worked in their family business early on in life. Good service and hospitality—meaning more than just rendering service—were foremost in the actions and deeds of the thirty women who participated in the study.

There were minimalist methods of inquiry used in the study. Potential participants in this research were contacted through verbal exchange after the purchase of a good or service, with interpersonal decorum creating a sense of trust between the two parties. Snowball methods of meeting someone based on another acquaintance was used as much as possible, again earning trust and understanding between researcher and interlocutor. Sometimes it would take more than one encounter to set up a time for conversation and more often that did not happen. During those moments when an interview was not possible, observation became the key component; just listening to the banter with others and watching interactions with consumer/visitors was a very valuable exercise. Being a fly on the wall was a solid ethnographic practice. Whenever possible, appointments were made and time allotments for interviews were kept to a minimum because these events took place in between paying customers or after work when it was a slow day or slow night depending on the job. These observations and informal interviews continued for years, as attempts to catch up during short visits, from reports from friends, and during the pandemic on WhatsApp.

Another goal of my research was to apply an intersectional lens to examine how women's choices or lack thereof were hindered by social obstacles such lack of education, skills, and what they thought of tourism as a way of making a living. Two things came to mind when drawing up a list of work categories. First was the fluidity of a job and then confluence of jobs. For example, among the four small hotel/cottage owners, all at one point in time served as housekeeper, accountants and more for their family businesses until they took over the establishments when their parents or grandparents retired. All the women workers benefitted from on the job (OTJ) training, with some skills requiring more diligence in the business on top of formal training. Finally, the professionals among the group earned their postsecondary education in Jamaica, in the United States, and Britain.

Depending on their age, many of these women tourist workers were leaders in their fields of endeavor. One was one of the leading housekeepers in a large resort, in charge of the duties of her co-workers and paid extra for this position. The table captain, by virtue of her job description, was the overseer of evening dining wait staff. Among the interlocutors were community leaders who were well established activists, particularly in the governance of Negril's tourist industry by way of the Negril Chamber of Commerce. A cadre of women were members of teams that are concerned with recycling, road clean up, and marine life management. Of the thirty participants in my study, two were native to Negril, but the majority commuted from nearby villages of Green Island, Lucea, and Sheffield. All actions were gendered, as well as classed, raced, and intersected with all other forms of difference.

Notions of respect and reciprocity were central in this work, as acts of Jamaicanness prevailed on my part and that of the women. It was evident that over the years when a return trip afforded a moment to catch up and to follow up with an interlocutor, the moment was welcomed.

Meanings of Women, Work, and Tourism

In Jamaica, forty-five percent of women are the primary providers (unicef.org). The historical record shows a precedence of Jamaican women being in charge of the economics of producing, providing or managing the resources essential to meeting the daily needs of their families. Children are welcomed and are the primary focus of a mother's work endeavors. For most women, there are also social obstacles to overcome, grounded in skin colorism, lack of access to resources including education beyond elementary level, that curtailed employment opportunities. Using an intersectional lens, the intertwining of race/color, class, and gender became clear. With sixty-two percent of the female work force (statin.jm) engaged in low skilled tourist work, it is evident that this sector opened -up jobs for poor and working-class women. After all job creation was the basic premise in tourist development in Jamaica and elsewhere in the Caribbean. UNWTO directives outline the importance of job creation as an avenue for women's empowerment. However, as a mature destination—Jamaica's tourism coming on stream full blown in the 1950s–1960s—the sector's solution to women's economic status and worth is an ongoing process. As evident in my research on tourism in Negril. Jamaican women's self-worth and ways of surviving with dignity are truly tested. To maintain that sense of self, skin-teeth are cultural practices often invoked as well as asserting one's Jamaicanness so the product—the satisfied guest who returns is the goal.

REFERENCES

Bolles, A. L. (1992). Sand, sea and the forbidden: Media images of race and gender in Jamaican tourism. *Transforming Anthropology* 3(1): 30–35.

Bolles, A. L. (2005). The golden goose: Race, gender and global tourism in Jamaica. In *Cultura y Desarrollo*. Special issue Culture and Diversity and Tourism. UNESCO Regional Office of Culture for Latin America and the Caribbean, no. 4, 82–94.

Bolles, A. L. (2022). *Women and tourist work in Jamaica: Seven miles of sandy beach*. Lexington Books.

Brennen, D. (2004). *What's love got to do with it*? Duke University Press.

Caribbean Tourism Organization. (n.d.). Caribbean Tourism Organization website. www.onecaribbean.org.

Carter, S., & Cannon, T. (1992). *Women as entrepreneurs.* Academic Press.

Dunn, H. S., & Dunn, L. L. (2002). *People and tourism*. Arawak Publications.

Enloe, C. (1990). *Beaches, bananas and bases*. University of California Press.

Fernandez, N. (1999). Back to the future? Women, race and tourism in Cuba. In K. Kempadoo (Ed.), *Sun, sex and gold: Tourism and sex work in the Caribbean*, 81–92. Rowman and Littlefield.

Freeman, C. (2014). *Entrepreneurial selves: Neoliberalism, respectability and the making of a Caribbean middle class*. Duke University Press.

Gibson, H. 2001. "Gender in Tourism: Theoretical perspectives." In Apostolopoulos, S. Sonmez, & T. J. Dalton (Eds.), *Women as producers and consumers of tourism in developing regions*, 19–43. Praeger.

Ghodsee, K. (2005). *The red Riviera.* Duke University Press.

Gordon, D. (1989). Women, work and social mobility in post-war Jamaica. In K. Hart (Ed.), *Women and the sexual division of labor in the Caribbean,* 67–80. The Consortium Graduate School of Social Science.

Hurston, Z. N. ([1938] 1981). *Tell my horse*. Turtle Island.

Kinnard, V., Kothari, U., & Hall, D. (1994). Tourism: Gender perspectives. In V. Kinnard & D. Hall (Eds.), *Tourism: A gender analysis*. Wiley.

Koulma-Castelberg, M. (1991). Greek women and tourism: Women's cooperatives as an alternative form of organization. In N. Redflift & M.T. Sinclair (Eds.), *Working women*, 197–212. Routledge.

Mintz, S. W. (1974). *Caribbean Transformations.* Johns Hopkins University Press.

Negril Guide. (2020.) *Little piece of paradise*. Negril Chamber of Commerce.

Nichols, G. (1992). *Daughters of Africa*. Ballantine Books.

Patullo, P. (1996). *Last resorts*. Ian Randle Publishers.

Rhiney, K. C. (2012). The Negril tourism industry: Growth, challenges and future prospects. *Caribbean Journal of Earth Science* 43: 25–34.

Roland, K. (2011). *Cuban color in tourism and La Lucha*. Oxford University Press.

Shepard, V. (1999). *Women in Caribbean history*. Ian Randle Publishers.

Statista.com/stastics/803809/Jamaica-gender-gap-labor-market/ category (accessed April 11, 2021).

Statistical Institute of Jamaica. statin.gov.jm (accessed October 29, 2019).

Stupart, C.& Shipley, R. (2103). Jamaica's tourism: Sun, sea and sand to cultural heritage." *Journal of Tourism Insights* 3(1): 4. https://scholarworks.gvsu.edu/jti/vol3/iss1/4.

Swain, M. (1995). Gender and tourism. *Annals of Tourism Research* 22(2): 247–266.

Taylor, F. F. (1991). *The hell with Paradise: A history of the Jamaica tourist industry.* University of Pittsburgh Press.

Timothy, D. J. (2001). Gender relations in tourism. In Y. Postolopaulos, S. Sonmez, & D. Timothy (Eds.), *Women as producers and consumers of tourism in developing regions,* 235–248. Praeger.

Tourism analytics. (2022). Jamaica. https://tourismanalytics.com/caribbean.html.

Trading Economics. (2021). Jamaica government debt to GDP. https://tradingeconomics.com/jamaica/government-debt-to-gdp
Travel Agent Central. (2021). *Caribbean.* Travelagent.com/Caribbean.
Ulysses, G. A. (2007). *Downtown ladies.* University of Chicago Press.
The United Nations Development Program. (2020). *Human development report.* New York.
UNWTO. (2021). Tourism Statistics database. https://www.unwto.org/tourism-statistics/tourism-statistics-database.
UNWTO (2022). History. https://www.unwto.org/history.
UNTWO World Tourism Barometer (2019). 17(2), UNWTO.
Visit Jamaica. (n.d.). Jamaica website. www.visitjamaica.com.
Visit Negril. (n.d.). The online hotel resource for travels to Negril. Jamaica website. www.visitnegril.com.
Williams, B. C. (2018). *The pursuit of happiness*. Duke University Press.
World Education News and Reviews. (2019). Education in Jamaica. September 10, 2019, www. Wcnr.wes.org/2019/09/education-in-Jamaica.

Chapter 5

Pedagogical Tourism

The Gendered Coloniality of Spanish Lessons in Guatemala

Sarah Becklake

INTRODUCTION

This chapter[1] focuses on "pedagogical tourism," a term used here to refer to diverse forms of tourism that claim to have and/or are undertaken for their educational potential. While tourism is often conceptually reduced to holidaymaking, it is better defined as a broader set of non-daily human spatial mobilities "characterized by choice, circularity, formalization, and consumption" (Becklake 2021: 209). Tourism, then, acts as an umbrella concept to include diverse forms of highly privileged non-daily human spatial mobility, including, but not limited to: holidaymaking, business travel, studying and volunteering abroad, conferences and field work, and visiting friends and family (Becklake 2021). Tourism is highly commodified and commodifying travel (Böröcz 1996). It has developed into one of the world's largest industries (World Bank 2017) and plays a central role in the function and spread of global capitalism (Fletcher 2011). From its roots in the European Grand Tour (Urry 2002), tourism has regularly been claimed to have "inherent" pedagogical properties. Indeed, Lyons et al. (2012) note that many policymakers, educators, and scholars regularly argue that tourism is supportive of cross-cultural understanding and cosmopolitan global citizenship. While such claims have and certainly should be critically interrogated, increasingly globalized and competitive labor and education markets have raised the strategic value of "global competence" (Brustein 2007). Following, those able to participate in international tourism are increasingly making the world their classroom. This implies a merging of the "tourist gaze" (Urry 2002) with an educational gaze, whereby tourists use immersive embodied experiences of difference as a means of learning about themselves and others (Doerr 2013, Vrasti 2013). Today a plethora of companies, universities, and non-profit organizations (Becklake 2014, Brustein 2007, Brown 2009, Caton and Almeida Santos 2009, Epprecht 2004) competitively offer study abroad programs, international internships, and volunteer placements, or a combination of all three. While touristic flows of students, interns, and volunteers occur within and between the Global North and the Global South, in this chapter I focus on flows from the Global North to the Global South and take a decolonial feminist approach. Following, I investigate not only how pedagogical tourism is shaped by and shaping of the gendered coloniality of power (Lugones 2010), but also how it is experienced and resisted by those living the "colonial difference" (Mignolo 2002).

Reflecting historically produced geo- and body-political inequalities (Grosfoguel 2009, Mignolo 2002, Quijano 2000), white Westerners continue to dominate the global tourism industry and have privileged access to international tourism (Becklake 2021, Mowforth and Munt 2009). Likewise, the white Western tourist gaze, which is shaped by (post-)colonial discourses that both exotify and problematize racialized "Others" (Hall 1992), powerfully governs places and people seeking to compete for international tourists (Becklake 2021). This is especially the case in the Global South, which is particularly dependent on international tourism as a tool of economic growth (Mowforth and Munt 2009). Within the Global North, travel to the Global South is deemed highly pedagogical. Indeed, Westerners use the geopolitical "colonial difference" (Mignolo 2002) as a resource to test and prove themselves, gain skills to improve their labor market positioning, and have life changing transformative experiences with "exotic" and/or "poor Others" (Crossley 2012, Vrasti 2013, Lozanski 2010). To facilitate this, Westerners can purchase a diverse array of touristic learning programs focused on ecology, indigeneity, and/or poverty (Becklake 2014, Caton and Almeida Santos 2009, Epprecht 2004), as especially seen in the case of volunteer tourism (voluntourism). While participating in pedagogical tourism in the Global South is highly celebrated in the Global North, various scholars have highlighted how these touristic learning programs, including study abroad and voluntourism, are not only shaped by geo- and body-political inequalities, but can (and often do) play a role in (re)producing them (Caton and Almeida Santos 2009, Epprecht 2004, Mostafanezhad 2013a, 2013b, Simpson 2004, 2005). While critical research has challenged the celebratory discourse surrounding pedagogical tourism in the Global South, it still tends to center the motivations and experiences of traveling Westerners (see, e.g., Caton and Almeida Santos

2009, Conran 2011, Crossley 2012, Kipp, Hawkins, and Gray 2021) to the exclusion of those living the "colonial difference" (Mignolo 2002).

As the above indicates, entanglements of tourism and education are highly shaped by and shaping of the continued "coloniality of power" (Quijano 2000). Anibal Quijano's (2000) concept of the "coloniality of power" denotes "a principle and strategy of control and domination" that is grounded in European racism and, thus, works to "justify" dehumanization and the violent occupation, appropriation, and exploitation of non-European places and peoples (Mignolo 2002: 83). While the coloniality of power was consolidated during direct colonial rule, it continues today. In particular, its contemporary manifestation is evident in and exercised through the continued body-political hierarchy between white Westerners and racialized Others; the production and domination of Eurocentric knowledge; and, the global capitalist system, which is built on colonialism/imperialism and, thus, characterized by geopolitical structural dependency (Bhambra 2021, Mignolo 2002, Quijano 2000). While the continued coloniality of power is clearly still evident in and exercised through symbolic, epistemic, and material hierarchical relations between "the West/Westerners" and "the Rest/Others" (Hall 1992), it also shapes the relationship *between* those living the colonial difference, as seen in the operation of imposed discourses and relations of race, class, and gender (Lugones 2010). Mignolo explains that decolonial thinking requires paying attention to "colonial legacies, colonial differences, and colonial wounds" (Mignolo 2018: 377). In advocating a gender-sensitive approach, Lugones calls "the analysis of racialized, capitalist, gender oppression 'the coloniality of gender'" and "the possibility of overcoming the coloniality of gender 'decolonial feminism'" (Lugones 2010: 747). Taking a coloniality of gender approach, as Lugones explains, requires focusing on entangled processes of oppression and everyday acts of resistance "at the fractured locus of the colonial difference" (Lugones 2010: 748). Taking inspiration from these decolonial scholars, this chapter asks: how are pedagogical forms of tourism produced and experienced by different actors in the Global South?

To address this question, the chapter focuses on study abroad language programs in Guatemala and investigates how Guatemalans living and/or working in the small city of La Antigua Guatemala (Antigua, for short), produce and experience providing Spanish lessons for—primarily white Western—language students. The data for the chapter comes from five months of ethnographic research[2] conducted in 2013 and was analyzed for diverse relations of power and inequality. While the imperial relation between Westerners and Guatemalans certainly shapes why and how Guatemalans produce and experience providing Spanish lessons in Antigua, race, gender, and class relations between Guatemalans also emerged as highly important. In particular, the chapter highlights how the Spanish "language tourism industry" (Tegelberg 2013) in Antigua is dominated by Ladinos—"people of mixed Spanish-indigenous heritage often said to reject Maya cultural values" (Tegelberg 2013: 82); how patriarchal gender relations and discourses construct Ladinas as the "right" kind of labor, leading to the feminization of teaching Spanish; and, how directors/owners of Spanish schools hold considerable power over Spanish teachers, affecting their strategies to benefit from and experiences of teaching Spanish. While the Spanish teachers I spoke with often expressed gratitude for and pride in their work, they also voiced anger and feelings of exploitation toward Spanish school directors/owners and their Western students. In short, the chapter exposes processes of exclusion, domination, and exploitation, as well as small acts of resistance. Like all research, there are limitations to the study. As the research was restricted to Antigua, it cannot account for differences that may be found throughout Guatemala. As all of the interviewees worked for officially recognized Spanish schools, the research is biased toward the formal sector. And, as the research took place in 2013, it cannot account for recent changes and events (e.g., the COVID-19 global pandemic and increased online language instruction).

The chapter proceeds as follows: first, it briefly introduces Antigua as a pedagogical tourism destination with a Ladino-dominated Spanish language tourism industry. Second, it outlines the feminization of teaching Spanish and the resulting "gender competitive advantage" (Elson, Grown, and Çagatay 2007). Third, it demonstrates how Spanish schools govern and control their teachers, creating challenging and sometimes abusive working conditions. Fourth, it turns its attention to how Spanish teachers aim to "take advantage" of their encounters with Westerners. Finally, fifth, it concludes.

ANTIGUA: A PEDAGOGICAL TOURISM DESTINATION

In 2014, the Guatemalan Tourism Institute (INGUAT) launched a new tourism marketing campaign called *Life Lessons*, with the slogan "*Guatemala: More than a journey, let Guatemala teach you life lessons.*" Through a series of short promotional videos, this savvy campaign promised would-be tourists that Guatemalan culture, nature, people, and wisdom will provide them with enjoyable and transformative educational experiences.[3] In promising "Life Lessons" in Guatemala, INGUAT is promoting the whole country as a pedagogical tourism destination and, thus, competing to attract international tourists looking to make the Global South their classroom. Rather than encourage new touristic discourses and practices, this campaign reflects and champions those already in existence. Indeed, Guatemala has long been a popular destination for Westerners seeking to engage with, learn from, and/or help indigenous Guatemalans (or "the Maya"[4]) (Burtner 2004, Chang 2017, Devine

2014, Little 2004, Tegelberg 2013), "natural paradise" (Cremers 2020), and "the global poor" (Becklake 2014, 2020, Vodopivec and Jaffe 2011, Vrasti 2013). While INGUAT's Life Lessons campaign places considerable emphasis on pedagogical experiences with Mayas and nature, in its promotional video entitled "True wealth goes beyond money," it states:

> Guatemala is the perfect place to come and learn Spanish. Not only will you learn the language, every teacher is eager to teach you something else; that is why, if you have come to learn Spanish, don't leave Guatemala without learning the life lessons it has to share with you.[5]

Despite the fact that Guatemala is actively being marketed as a Spanish-learning destination, and that Spanish schools and teachers have become key tourism service providers and workers, few studies have focused on the country's language tourism industry (notable exceptions are: Tegelberg 2013, Barrera Nuñez 2009). In picking up this gap, here focus is placed on the city of Antigua, Guatemala's main tourism destination and home to a plethora of Spanish schools.

The construction of Guatemala—and, by extension Antigua—as "the perfect place to come and learn Spanish" has its roots in Spanish colonialism (starting in the early 1500s), which resulted in the dominance of Spanish culture, including the language, and an enduring hierarchical relationship between the Ladino and Maya populations (Adams 2005, Quiroa 2013). While Guatemala's long and bloody civil war (which resulted in genocide against the Maya) ended in 1996, the country remains beset with high levels of inequality, poverty, violence, and crime, and the Maya are overrepresented amongst the poor (Sanchez, Scott, and Lopez 2015). Despite its immense socio-economic problems, international tourism has become highly economically important for the country, especially in tourism-dependent Antigua (Becklake 2021, Burtner 2004, Little 2014). While tourists expect and often enjoy encounters with Mayas in Antigua (Little 2004), the city's hotels, restaurants, shops, and Spanish schools are "primarily owned and staffed by Ladinos that cater to international tourists" (Little 2014: 400–401). As anthropologist Walter Little notes: "for Antigua's political and business leaders Antigua is not a Maya place. Rather, it is a Ladino and Spanish colonial place" (2005: 86). While the Ladino-dominated city draws heavily on and benefits from Maya culture and labor (Little 2004), indigenous vendors are generally "not welcomed by the local authorities and non-Maya businesspeople" (Little 2005: 94), who complain "that street vendors disrupt their businesses and taint the colonial image" (Little 2003: 530). Mayas, thus, often experience exclusion, discrimination, and criminalization in Antigua (Becklake 2020, 2021, Little 2014, 2009). Reflecting the fact that Guatemalan

Figure 5.1. Antigua's most touristic street, Calle del Arco.
Source: Photo taken by Sarah Becklake, 2013.

tourism branding/marketing has placed emphasis on the "exotic," "colorful," and "mysterious" Maya since the 1930s (Little 2008), most studies of tourism in Guatemala focus on the Maya. Indeed, the "exotic/poor" Maya are highly attractive not only for Western vacationers and volunteers, but also for researchers (Barrera Nuñez 2009). While in other parts of Guatemala, the Spanish language tourism industry includes more Mayas as directors/owners and teachers (Tegelberg 2013, Barrera Nuñez 2009), in colonial Antigua it is largely a Ladino-dominated business.

Once the Spanish capital in Guatemala, Antigua is characterized by colonial architecture and ruins, its "Eternal Spring" climate, and the three volcanoes surrounding it. Made a UNESCO World Heritage Site in 1979, tourists are drawn to Antigua for its beauty and history, but also for its accessibility to the international airport, high levels of comfort and services, and as a "safe place" to have touristic encounters with Mayas, to volunteer, and/or to learn Spanish (Becklake 2014, 2020, Little 2004, 2014). The popularity of Antigua as a place to learn Spanish is often explained by its touristic charm, services, and relative security, *and* the plethora of Spanish schools, their specialization in affordable yet high quality one-on-one (one teacher for one student) Spanish lessons and the clear Antigüeñan accent. According to its website, Fundación Proyecto Lingüístico Francisco Marroquín (PLFM) was Antigua's first Spanish school, opening in 1969 during Guatemala's 36-year civil war (1960–1996). Interestingly, PLFM claims to be "run by native Mayan professionals" and has two key purposes: to teach the Spanish language and "to teach, investigate and preserve the Mayan languages spoken around Guatemala and the [c]ulture."[6] Following PLFM's success (it is still in operation today), many Ladino-run Spanish schools have since opened in the city. In 2013, INGUAT's website noted 19 registered Spanish schools in Antigua, yet interviewees estimated between 35 and 60 formal and informal schools. While some Spanish schools are set up as non-profit organizations or cooperatives, others are small businesses. Many of Antigua's Spanish schools offer comprehensive touristic learning packages, including one-on-one Spanish lessons, homestays with Guatemalan families, volunteer opportunities at local non-governmental development organizations (NGOs), numerous fun touristic events and activities, and encounters with Mayas (on the latter, see Little 2019). Indeed, Spanish schools effectively operate as mini tour operators and language students in Guatemala are often voluntourists and vice versa (Barrera Nuñez 2009, Tegelberg 2013, Vrasti 2013).

Today, many of Antigua's formal Spanish schools are registered with INGUAT and the Guatemalan Ministry of Education, are affiliated with Guatemalan universities, and some are accredited by universities in the Global North, allowing large groups of Western university students to come to Antigua to take Spanish lessons for credits back home. Primarily from Canada, Europe, and the United States,[7] thousands of high school and university students, missionaries, volunteers, professionals, soldiers, and retirees come to Antigua every year to study Spanish. While socio-demographic statistics of these language students could not be found, research on US study abroad participants show that they are predominantly middle-class, white, and female (Chang 2017). In their ability to attract thousands of Westerners to Antigua every year, Spanish schools have become powerful tourism actors. They help support tourism businesses in Antigua and are key sources of work for many Guatemalans. Indeed, their comprehensive touristic learning packages depend on the labor of Guatemalans working as Spanish teachers, homestay providers, event and volunteer coordinators, and Maya cultural performers. In this chapter, I focus on Spanish teachers. One Spanish school director estimated that there are upward of 2,000 Spanish teachers in Antigua alone. In a city with an estimated overall population of 45,000 (Little 2014), teaching Spanish clearly represents a significant source of employment. To get a job as a Spanish teacher at one of the formal schools, Guatemalans must take time-consuming teaching courses, pass teaching exams, demonstrate considerable knowledge of Guatemalan culture and history, go through difficult selection processes and interviews, and pass strict probation times. While the amount of economic and cultural capital needed to become a Spanish teacher is prohibitive for many, it is especially prohibiting for indigenous Guatemalans; although, as already noted, some Spanish schools in Antigua do offer instruction in Maya languages and, thus, hire a few Mayas.

The increased value placed on global competence, the notion that travel from the Global North to the Global South is especially pedagogical and worthy of celebration, and the growing demand for Spanish-language instruction (International Association of Language Centres 2016), have all helped to produce significant Western demand for Antigüeñan Spanish schools' comprehensive touristic learning packages. In the process, the white Western tourist/education gaze takes on considerable governing power in Antigua (Becklake 2021). Indeed, in their aims to attract and satisfy international language students, Spanish school directors/owners engage in "tourism reflexivity" (Urry and Larsen 2011); that is, they learn to see Guatemala and Guatemalans through the white Western tourist/educational gaze and, thus, come to know their competitive (dis)advantages as a pedagogical "tourism destination" and "touristic figures" (Becklake 2016, 2020, 2021). Following, they engage in strategies to represent and govern their school and teachers in ways that meet Westerners' needs, expectations, and desires. As Epprecht (2004) notes, North American students often have a sense of entitlement; they see themselves as consumers of education and, thus, often expect their "First World" needs be met even when in "Third World" contexts. Similarly, my research suggests that Spanish school directors/owners and teachers believe that Western language students want "First World" quality at "Third World" prices. In aiming to provide high quality, but cheap Spanish lessons, Spanish school directors/owners place considerable emphasis on who they hire as teachers and how they monitor and control them. This, in turn, deeply shapes teachers' working conditions and how they experience their jobs. In what follows, I show how relations of gender and class further shape how Ladino Spanish school directors/owners and teachers produce and experience providing high-quality, yet affordable, language lessons for white Westerners in Antigua.

"IT'S A WOMAN'S JOB": THE FEMINIZATION OF TEACHING SPANISH

While the majority of Spanish school directors/owners are men, the vast majority of teachers in Antigua are women. While one Spanish school director estimated that approximately 70 percent of all Spanish teachers are women, at another school I was told that 85 percent of their 55 teachers are women. Another two schools said they only hire *maestras* (female teachers). Given that Guatemala has high levels of poverty and a "lack of opportunities" for women in the labor market (Eberhard-Ruiz 2021: 14–15), teaching Spanish undoubtably represents a compelling option for educated women. Spanish school directors, owners, and teachers were all willing to discuss the reasons why teaching Spanish is a "woman's job." Their answers echoed well-trodden discourses associated with heterosexual patriarchal relations and a gender division of labor in the home, whereby men are deemed husbands/fathers responsible for earning a family wage and women are deemed wives/mothers responsible for domestic work and childcare (on gender relations see, e.g., Enloe 2014). While actual lives rarely neatly conform to these gendered discourses, interviewees used them to explain who chooses (labor supply), but also who is chosen (labor demand) to become Spanish teachers.

Given the fluctuation in demand for Spanish lessons, Spanish schools rarely hire teachers as permanent employees but, rather, as flexible freelancers without job security or benefits. Likewise, even if they work at a formal school, many Spanish teachers are part of Guatemala's informal labor force. Whether one receives teaching hours depends on the teacher's internal hierarchical ranking (see next section) within the school and the school's current number of students, so teachers never know when they might be given teaching hours. Despite the fact that Guatemalan men dominate in other highly precarious tourism jobs, such as tour guides (LaPan et al. 2021), Spanish school directors, owners, and teachers claimed men cannot choose to become Spanish teachers because they should normally work full-time. Conversely, they argued that teaching Spanish works well for women, who, apparently, are only looking to supplement their husband's earnings and need flexible part-time work that combines with their primary responsibility for the home and children.

> The other reason [I only hire women] is because the majority of our students come for four hours and I think that the work of a Spanish teacher is an option for women and that many of them cannot work full-time because they have children, they have other responsibilities, so four hours for them is good. In contrast, normally a man should work full-time. (Spanish school owner/director, male)

> The Spanish classes are for the wives. Because for the wives, it is like *un dinero extra* [extra money]. It is very flexible work, because if your child is sick, you can cancel the class and reschedule it for Saturday . . . And this is one of the reasons why there are so many women, because it is a job for the women, for the wives, who are responsible for everything in the house. A husband could not be a Spanish teacher . . . No, it is very unstable work. (former Spanish teacher now part-owner of Spanish school, female)

While many of the teachers I spoke with expressed gratitude for paid work that they can combine with their domestic and caring responsibilities, the idea that Guatemalan women can somehow "afford" precarious paid work is short-sighted. As wages in Guatemala are generally low (Eberhard-Ruiz 2021), husbands' earnings alone are often not enough to support families. Furthermore, many Spanish teachers are university students and/or single mothers. In the latter case, their wage may be the only one supporting a whole household. Women's supposed tolerance of flexible working conditions supports "organizational flexibility" (Holtgrewe 2007), which allows companies to respond quickly to fluctuations in demand with the least cost (Sinclair 1997). Yet organizational flexibility also makes private lives highly inflexible (Peterson and Wiens-Tuers 2014). Thus, while teachers can in principle choose which teaching hours they take or not—allowing them to take hours that fit around other responsibilities—various teachers complained that they can rarely plan their lives in advance, as they never know when they will be called to work. Evelyn, a teacher with 35 years of experience, explained that when the school calls one should take the teaching hours lest the school thinks you do not want to work. However, this can be very challenging. As teaching hours are usually given in week-long blocks, taking teaching hours when they are given means having to cancel anything that was booked in advance. As Evelyn explains, this is particularly difficult with hard-to-get medical appointments.

While there are some male Spanish teachers, many interviewees emphasized that Spanish schools prefer to hire women because they are more responsible and professional. Indeed, while "responsible fathers" are thought not to be able to choose to become Spanish teachers, Guatemalan men are also thought to be too irresponsible to teach Spanish. Interviewees suggested that male teachers are more likely to arrive at the school hungover and smelling of alcohol. Conversely, women are argued to be more reliable because of their gendered responsibilities. As Beatrice, a former Spanish teacher who has since become part-owner of a small Spanish school, explained:

> The first reason [why there are more women than men as teachers] . . . is that sometimes the men are very irresponsible, and more with this type of work. . . . And, so, I believe that in this sense they [directors] know that women have more responsibility and sometimes they have their children and so they know that they are never going to fail . . . I am a woman and I know that women do not fail you . . . I think that right now the directors have realised that women are more secure; they are more responsible.

While professionalism is often linked to teachers' certification and experience, it is also symbolized by how one comports and presents oneself (Waring and Waring 2009). In not arriving at the school hungover and smelling of alcohol, women are also deemed more professional. Yet rather than merely capitalizing on Guatemalan women's "inherent" professionalism, Spanish schools play a key role in producing it. Indeed, many schools have uniforms and/or strict rules surrounding dress and appearance, requiring their teachers to perform considerable "aesthetic labour" (Witz, Warhurst, and Nickson 2003). Evelyn explained how she puts in a lot of money, time, and care into how she presents herself: shopping for second-hand clothing at the local market, showering before every class, and keeping her hair well-trimmed and tidy. All of this "body work" (Wolkowitz 2006) disguises her relative poverty from her students. Indeed, the day I visited Evelyn in her humble home she looked considerably different. Spanish teachers' body work reduces perceived inequalities between students and teachers, thus helping make students feel more comfortable.

Looking and acting professional is also interpreted as containing "undesirable" expressions of (hetero)sexuality. Guatemalan men, it is suggested, are not to be trusted as Spanish teachers, as they flirt with female students and try to start sexual relationships with them. This evokes colonial tropes of predatory "Othered" men, who have no control over their sexual urges, unleashing unwanted sexual attention on white women who need protection from them (Becklake 2021). Conversely, Guatemalan women are deemed sexually unthreatening and/or attracting. However, how they do or do not express their sexuality at work is also highly regulated. As one male director stated: "We don't have uniforms because I don't want them to all look the same, however everyday everyone has to dress decently, not too sexy." While women are hired because they are deemed sexually unthreatening, they are also expected to perform considerable "sexual labour" (Adkins 1995) for the benefit of both male directors and (presumably male) students. One teacher explained how a director would only hire women he found attractive. He would try to start sexual relationships with the teachers and if a teacher did not reciprocate they would not receive teaching hours. Beatrice suggests that some directors also hire women who are open to "adventures with students." Likewise, while women are hired for being "respectable," becoming a Spanish teacher can call their respectability into question. As Beatrice explained, especially "in the past, if you said, 'She is a Spanish teacher,' very quickly the whole world is thinking that she has 'adventures' with the students." For this reason, some husbands are not supportive of their wives becoming Spanish teachers. Carlos, my Spanish teacher for one week, told me that he would not want his wife teaching Spanish because he knows "what goes on," and, as he implied, participates in it himself.

Women are also deemed "naturally" good at teaching Spanish. When I asked Daniel, a school director, what makes a good Spanish teacher, he replied, "Personality. That's the main thing." Many interviewees suggested that women inherently have the "right" personality to teach Spanish, including being talkative and creative, as well as patient and caring. Beatrice explained:

> Look, we [women] like to talk . . . We love to talk and women always have conversation topics and everything. I have heard some male teachers in class. What are men's topics? Football, women, and drinking. I am not going to say all male teachers . . . there are male teachers who are very cultured, but the majority of the male teachers do this . . . So, for me, women have many conversation topics. And, so, this helps you a lot. We love it.

Yet rather than being "naturally" good at making conversation, Evelyn told me how she works hard to build her cultural capital (e.g., by reading the newspaper daily) in order to be able to converse on many different themes. Spanish teachers also put considerable effort into how they talk, consciously speaking slowly and choosing words and tenses carefully. This may be particularly important in a destination which is, in part, sold for having a clear and exact way of speaking Spanish. Beyond the normal remit of a teacher, Guatemalan Spanish teachers also listen to their Western students' problems and experiences of culture shock in the "Third World" and try to give them advice and help them to adapt. One teacher noted that, "There are many students who come with psychological problems." The youth groups she is talking about come to Guatemala to have "life changing experiences" and she becomes part of that experience. She explained that she treats them like her own children and gives them advice. When I remarked that it sounds like she acts as a mother to them, she stated: "Yes, exactly."

Feminist scholars have shown how gender segregated labor markets, the assumption that women's wages are only supplementary to men's, and the "naturalization" of women's labor and skills all help to construct women as "cheap" workers (Enloe 2014). Interestingly, teaching Spanish *can* be a relatively well-paid job for women in Guatemala. Depending on teaching hours, one director suggested that teachers can earn between Q4,000 and Q5,000 (US$526 to US$657[8]) a month, far more than the 2013 minimum monthly salary of Q2,171[9] (Guatemalan Ministry for Work & Social Security n.d.). However, another director estimated that, on average, teachers earn between Q2,000 and Q3,000 (US$256 to US$385) a month and, thus, they often earn close to the minimum monthly salary. Yet, given the fluctuation in teaching hours, these wages are also highly unstable. Weeks or months can go by without work, and, thus, Spanish teachers' annual wages are likely much lower than those earning the minimum monthly salary. This is not a living wage in or around touristy Antigua. When I asked one teacher if her salary was enough to live off she replied: "You could say that it is only a way to live a little, yes, because if you use only it, no, no." Likewise, almost all of the teachers I spoke with had other jobs to "make do" (Elson 1992). Rather than pulling double-shifts, many of these women do triple-shifts or more as a means of substituting their precarious wages as teachers. Feminist economists argue that the construction of women's labor as "flexible" and "cheap" allows it to be strategically used by export-oriented companies to keep costs low, leading to "gendered competitive advantages" in the global economy (Elson, Grown, and Çagatay 2007). While undoubtedly the affordability of Spanish lessons in Antigua for Westerners in part lies with global exchange rates, the lower relative cost of living, and the intense competition between Spanish schools, it is also linked to the feminization of teaching Spanish. In other words, the construction and exploitation of Guatemalan women as "flexible-cheap-professional" labor has helped to create Antigua's gendered competitive advantage as a Spanish-learning destination.

"IT'S LIKE A BIG COMPETITION": THE HIERARCHICAL RANKING OF TEACHERS

Most Spanish schools in Antigua, I was told, allocate teaching hours via a system of weekly rotation. As Evelyn explained, this system is fair "because . . . with this system of rotation everyone is able to eat a little." But rather than all teachers moving smoothly through the rotating list, teachers are hierarchically ranked by Spanish schools and those with high rankings receive more teaching hours. School director Daniel informed me that once a month their teachers are (re)ranked 1 to 50 and that the list is made public at the school. He asserted that "we are not cruel" but rather "it is like a big competition." There are at least three ways that teachers move up or down the hierarchical ranking: student evaluations, showing dedication to the school, and compliance. These disciplinary tactics work not only to help create "high quality" one-on-one lessons for Western students, but also create—at least on the surface—a docile workforce that constantly fears losing its ranking and, thus, its already precarious income.

As is common practice in many universities around the world, Antigüeñan Spanish schools use student evaluations of teachers as a method of monitoring and improving the quality of teaching; it is part of the process of professionalization. But unlike most universities, Guatemalan Spanish teachers are evaluated far more frequently (e.g., on a weekly or biweekly basis); are evaluated not only on teaching and class contents, but also on personality and appearance; and, the consequences of poor evaluations are drastic. Indeed, teachers who receive poor evaluations are punished by receiving fewer or no teaching hours, losing a substantial part of their already precarious incomes. These evaluations, then, hold considerable disciplinary power (Foucault 1977). Indeed, teachers work hard to determine each student's unique wants, wishes, likes, and dislikes, and often expressed pride in their ability to accommodate them. As one director stated, "The teachers put in a lot of effort so that they get along well with their students, whichever personality." Getting along well requires being 100 percent flexible. As one teacher explained, "We have to be willing to walk, to teach, to travel, to speak, to do the things that the student likes." Teachers also learn to censor their own ideas, opinions, and feelings, which requires considerable amounts of "emotional labor" (Hochschild 1983). For example, one day I remarked to Evelyn that my male Spanish teacher told me a "joke" about domestic violence in Guatemala. She responded with shock, not because of the inappropriateness of the joke, but because he works with foreigners and should have known that foreigners do not accept those things and, thus, should have been better at filtering what he says. In playing to Western ways of seeing/thinking, language students are often spared confrontation with cultural differences, suggesting that these one-on-one encounters lose much of their pedagogical potential.

Teachers may also move up and down the hierarchical ranking according to the dedication that they show to their jobs and the school. Indeed, teachers who show dedication are seen favorably in the eyes of directors, who ultimately make the decision regarding who does and who does not work. School director Daniel informed me that his school provides free training for the teachers on Saturdays. To Daniel coming to these trainings demonstrates that the teachers "want to improve" and, thus, it shows "good attitude." Another way that teachers show good attitude is through engaging in school-arranged activities and trips with their students. As Daniel explained, engaging in these activities demonstrates the teacher's dedication to the school:

> If you are a teacher and if you do a good job as a teacher, that's okay, but we also want to see what else you can do for your company. I mean, that's support. So, we have some teachers, like, for example, we are organising a trip to the Acatenango volcano next week, and there are 25 teachers that are going there. I mean they are having fun, but they are participating in the organisation.

Indeed, traveling with students is seen as part of a Spanish teacher's duties. While this enables some teachers to see places in their own country that they would otherwise not be able to afford to visit, Evelyn also explained that teachers cannot really say "no" if their students want to travel. If they say "no" they will have problems with the school; if they say "yes" they are demonstrating collaboration. While these trips are free for teachers, they also require teachers to spend more time with students than they are paid for. Taking Saturdays to do trainings, engaging in extra activities, and traveling with students—conceptualized as "showing dedication"—may be especially difficult for the largely female staff. When I suggested this to Daniel he agreed, but argued that things need to change. He argued that teachers who do not have supportive partners should re-think if they are in the right relationship. While Daniel's school benefits from "flexible-cheap-professional" female labor, he also challenges the status quo of feminized responsibility for children and male dominance over women in the household. As he stated: "I mean, they cannot be like in the past, 'Okay I do whatever you want me to do.' It is not like that anymore Sarah." While it is certainly interesting to consider how the feminization of teaching Spanish may help challenge unequal gender relations, Daniel also assumes that there is a man at home to help, forgetting, it would seem, that many Spanish teachers are single mothers. Consequently, "showing dedication"—and, thus, moving up the teacher hierarchy—is likely far easier for the few male teachers than the female majority.

Finally, teachers who are non-compliant quickly lose teaching hours. Coming to be seen as "problematic" is not hard to do. As outlined in the next section, there are many rules regarding what Spanish teachers are and are not allowed to talk about and do with their students. Evelyn explained that not taking teaching hours when offered, asking for time off (e.g., to go to a doctor's appointment or care for a sick child), or declining to travel with students are all frowned upon. Raquel, a Spanish teacher who has worked at numerous different schools, suggested that to not be seen as "problematic" requires being accepting and submissive. As she stated, "If you say something that they do not like, they will simply not give you students because they do not have any commitment to you." This means accepting exploitative and sometimes abusive working conditions. Indeed, the teachers I spoke with recounted stories of sexual and verbal harassment, bullying, of directors not paying wages on time or not at all, and of some schools putting limits on how often teachers can use the toilet. But not all schools are the same. Some schools have been set up as responses to the poor working conditions in others. This was the case of Beatrice, who emphasized that, "The directors do not make the school, the teachers make the school, and for this reason we need to support the teachers." By extension, one could say that Spanish teachers make the Spanish-learning destination. Yet, rather than receiving support from Spanish schools, most of the teachers I spoke with tended to complain about their working conditions and treatment. In this context, some try to build strategic alliances with their students. Indeed, as intimate embodied encounters, one-on-one Spanish classes also hold the potential for building familiarity, something Spanish teachers can use strategically to make teaching Spanish work for them. I highlight this in more detail below.

"TAKING ADVANTAGE": THE (MIS)USE OF IMPERIAL TOURISTIC ENCOUNTERS

Working as a Spanish teacher entails participating in face-to-face "imperial encounters" (Doty 1996). Scholars studying imperial touristic encounters in the Global South have highlighted that they are not only exploitative, but can also represent strategic opportunities to build intimate (platonic and sexual) relationships that can result in financial support and opportunities for travel and migration (Brennan 2007, Cabezas 2009, Kempadoo 1999, Simoni 2014). Likewise, while Guatemalan Spanish teachers labor under highly challenging and exploitative conditions, their encounters with Westerners create strategic opportunities to gain cultural, social, economic, and mobility capitals to which they would otherwise not have access. Nevertheless, as I show here, Spanish schools aim to govern and control if or how Spanish teachers are able to "take advantage" of their encounters with their Western students.

While many Westerners come to Antigua for pedagogical touristic experiences, Spanish teachers also learn from their encounters with Westerners. As Beatrice explained: "I have never left Guatemala. . . . So, with students I learn tradition, histories; they describe their cities to you and, in your imagination, you can travel to their places." Spanish teachers also learn about different ways of thinking and living and develop their own cultural and national stereotypes, with key differentiations made among students from the United States, Europe, and Asia. Maria, a former Spanish teacher who now works as a host-mum, stated: "We have realised there are some things that are very different . . . their ways of being, their attitudes to various things, for us they are different." In aiming to accommodate difference, the Guatemalan women I interviewed stressed that they have become cosmopolitan in their "attitudes, practice and abilities" (Vertovec 2010: 64). Being a Spanish teacher also creates opportunities for Guatemalans to build more intimate relationships with students. While stories of romantic relationships resulting in financial support and/or travel abroad are plentiful, more generally friendly relations with students allows teachers to create cross-border social networks. Such networks, as Simoni (2014) suggests, can provide a security net for people in the Global South, who often have little else to fall back on in times of economic trouble. Spanish schools, however, try to regulate how teachers benefit from their intimate encounters with students. For example, in some schools, students are automatically given a new teacher every week. While some argue that there are pedagogical reasons for this practice—listening to a new accent and experiencing a new teaching style—others suggest that it actually wastes the students' time and money. When I asked why this practice exists, Beatrice explained that Spanish schools do it "because they do not want them [teachers and students] to build a relationship; a familiarity." I ask Beatrice about the danger of such a familiarity, and she stated, "I think that what they want to avoid is that there is a lot of familiarity and that, in the end, your teacher is your friend and, then, maybe they take advantage of the student."

Within Antigua, there is a moral distinction made between different ways of benefiting from imperial touristic encounters. While transparent economic transactions, such as selling tourism services or products, are deemed "fair" and "correct, "incorrect" ways include guilting or duping tourists out of money or starting disingenuous intimate relationships with them. When this happens, Guatemalans are said to be "taking advantage" of tourists. The discourse of "taking advantage" positions Guatemalans as "strategic predators" and tourists as "naive victims," and creates a situation in which Guatemalans must constantly prove their good intentions and tourists are always on guard (in the context of Cuba, see Simoni 2014). Spanish schools put rules in place to try to hinder teachers from "taking advantage" of their students. For example, in some schools teachers are discouraged to talk about their economic hardships with students; they are forbidden to receive tips/gifts from students; they are not allowed to have any contact with students outside of formal teaching hours; they are not supposed to maintain contact with students after their classes end; they are supposed to contain classes to school premises unless otherwise authorized; and, they are not allowed to receive commission from stores for bringing them customers. While done in the name of protecting students from imagined "predatory teachers," such rules may actually primarily aim to protect Spanish schools' reputations and middle-men functions. Indeed, creating more intimate relationships may lead students to find out about their teachers' poor working conditions and/or lead students and teachers to meet for private Spanish classes outside of schools. Given the number of stories of more meaningful relationships that have developed between students and teachers, it is debatable whether such rules are successful. Rather than stopping deeper connections, these rules force teachers and students to act in a clandestine manner.

But who is taking advantage of whom? While Westerners often come to Guatemala to learn Spanish *and* help the "global poor," their behavior is often far from enlightened and supportive. Chang's (2017) study of US study abroad students in Guatemala explains how the minority Latina participants disparaged at the majority white participants' behavior, which included: clumping together, exotifying and commodifying indigenous people, engaging in condescending acts of charity, showing little respect for sacred places, making derogatory comments about indigenous culture, and unfairly bargaining with vendors. Similarly, despite hours of one-on-one face-to-face contact with Spanish teachers, many students never come to think about the conditions under which their teachers work and the economic precarity they face. They also, as Beatrice suggested, do not generally seem to care. She stated:

> Look, in all of my experience I remember two people, two students, in one school where I was during this time, and this person was charged 110 dollars, now I think they are charging 150, 160 dollars. And they said to me: "Well, I came to this school for the price . . . because I know that if there is a higher price the teachers are going to earn well." So, I said: "there goes an extraordinary person who thinks about the workers."

In fact, rather than express concern for their teachers' wages, a more common practice amongst students is walking around Antigua and bargaining for the best deal. As Beatrice explained:

> They come here and they know that here they can find even cheaper [Spanish classes], cheaper, cheaper. So, they go looking for the cheapest. And, so, it is like I told you, what salary can you give a person? How are the people going to live?

Bargaining is a common Western practice in the Global South. As studies from Guatemala and other Latin American countries demonstrate, Westerners often worry they are seen as walking dollar signs and, thus, asked to pay "*gringo* prices" (Schneider 2018, Little 2004). While bargaining may be more-or-less expected by both parties in some contexts (e.g., souvenir markets), Spanish teachers, who are aware of what students would have to pay for the same service in their own countries, see bargaining for cheaper Spanish lessons as highly insulting. They know they are already the cheap option and struggling to survive with their precarious wages. In this context, Spanish teachers may indeed try to find various ways of benefiting from their intimate encounters with Western students. In doing so, however, they are demonized for "taking advantage" of students, while bargaining students are often not given the same moral reproach.

Beatrice wishes students were more aware of their teachers' struggles. Yet, as Spanish schools prohibit teachers from discussing such topics with students, creating this awareness is challenging. The combination of not having their struggles recognized and constantly listening to their Western students explain their desires to "help" Guatemalans and seeing donations flow toward development NGOs, can create considerable resentment. This was eloquently expressed by Beatrice:

> Supposedly many foreigners come because they have the desire to help. So, where is the help? I do not know where the help is . . . Even though they [volunteer groups] go to the communities and help and everything, they do not realize that the Spanish teachers also need collaboration. Don't the Spanish teachers have children? They bring bags of books to the communities and what happens with the Spanish teachers? They do not stop to think that they are also there. Because we say we are surviving with this salary, not living, surviving. And, so, we said "Well, give bags of books to the teachers too." And they did not do it.

The vast majority of Westerners coming to Guatemala to "help" want to work with the Maya (Becklake 2014, 2020), who face higher rates of extreme poverty (Sanchez, Scott, and Lopez 2015). While Western language students may not see their Spanish teachers as part of the deserving global poor, Guatemalan Spanish teachers may come to see themselves that way. This was clearly expressed through my long interview with Maria, who emotionally explained how constantly listening to her Western students tell stories of their lives and lifestyles in the Global North and witnessing their culture shock to "Third World" Guatemala makes her feel poor. While Maria would likely be categorized as middle-class in Guatemalan terms—indeed, she and her family live in the center of Antigua—through her constant encounters with Western language students she comes to identify with "the colonial difference" (Mignolo 2002).

CONCLUSION

Within the Global North, a celebratory discourse surrounds pedagogical tourism in the Global South. This discourse is supported by a dearth of empirical research focusing on how touristic learning programs are produced and experienced by people in the Global South. Taking inspiration from decolonial scholars, this chapter has focused on how Guatemalans produce and experience providing Spanish lessons for Westerners in Antigua. In doing so, it highlighted four intersecting power relations—the imperial/race relation between Westerners and Guatemalans and Ladinos and Mayas; the gender relation between Ladino men and women; and, the class relation between Spanish school owners/directors and Spanish teachers—and exposed processes of exclusion, domination, and exploitation, as well as small acts of resistance. Spanish school directors/owners compete to attract/satisfy international—primarily white Western—language students *and*, in the process, create work for many Spanish teachers. This positions Spanish school owners/directors as powerful middlemen bringing Western students and Guatemalan Spanish teachers together. In their aim to give Western language students what they want—high quality, but cheap one-on-one Spanish lessons—they strategically hire, regulate, and discipline (primarily) Ladina women as "flexible-cheap-professional" teachers, helping to produce Antigua's gender competitive advantage as a Spanish-learning destination. While the Guatemalan women I spoke with often expressed appreciation for and pride in their teaching jobs, they also explained how they are exploited and controlled. Built on the exploitation of Guatemalan women, Antigua's gendered competitive advantage as a pedagogical tourism destination benefits Western students and universities, who have effectively outsourced teaching Spanish to these schools, as well as the many Spanish school owners/directors, touristic businesses, and development NGOs who benefit from the steady influx of Western language students to Antigua.

Pointing out the exploitation of women's "flexible-cheap-professional" labor in the Global South for the primary benefit of consumers from the Global North is not new (Enloe 2014, van Straveren et al. 2007). However, unlike export-oriented industries that send products produced in the Global South to the Global North, tourism brings Western consumers from the Global North to the Global South for immersive experiences and embodied encounters with the colonial difference. Working as a Spanish teacher, then, entails continual participation in and experiences of one-on-one imperial touristic encounters. These encounters are highly regulated by Spanish schools. As the chapter has demonstrated, this affects what students can and do learn, but also how Guatemalan women experience teaching Spanish. While Westerners come to Guatemala to have "life lessons" in the Global South, in playing to their ways of seeing/thinking and regulating what can/cannot be discussed, Westerners' confrontations with the colonial difference are curated and curtailed. Indeed, they are never asked to confront the continued coloniality of power and the exploitative relations that shape the production of pedagogical tourism in the Global South. Conversely, Spanish teachers cannot escape learning about and being confronted with living the colonial difference. They are continually reminded that they are part of the global poor, but apparently not poor enough to warrant Western "help." Yet these encounters also represent strategic opportunities for Guatemalan women, allowing them to gain access to a range of capitals that they would otherwise not have access. Through subversively building more meaningful intimate connections with their students, Spanish teachers resist and challenge the power of Spanish school directors/owners and aim to make these imperial touristic encounters work for them.

REFERENCES

Adams, R.N. (2005). The evolution of racism in Guatemala: Hegemony, science, and antihegemony. *Histories of Anthropology Annual* 1:132–180.

Adkins, L. (1995). *Gendered work: Sexuality, family and the labour market*. Open University Press.

Barrera Nuñez, J.O. (2009). Desires and imagination: The economy of humanitarianism in Guatemala," In W.E. Little & T.J. Smith (Eds.). *Mayas in postwar Guatemala: Harvest of violence revisited.* 110–123. University of Alabama.

Becklake, S. (2014). NGOs and the making of "development tourism destinations": The case of "Destino Guatemala." *Zeitschrift für tourismuswissenschaft* 6(2), 223–242.

Becklake, S. (2016). "Making Destino Guatemala": Everyday enactments of global tourism competition in La Antigua Guatemala. PhD diss., University of Lancaster.

Backlake, S. (2020). The role of NGOs in tourist securitization: the case of La Antigua Guatemala. *Space and culture* 23(1), 34–47.

Becklake, S. (2021). Genderd touristic security: The case of Guatemala. In M. Mostafanezhad, M. Córdoba Azcárate, & R. Norum (Eds.). *Tourism geopolitics: Assemblages of infrastructure, affect, and imagination.* 207–230. University of Arizona Press.

Bell, E.R. (2012). "Pirates of our spirituality": The 2012 apocalypse and the value of heritage in Guatemala. *Latin American Perspectives* 39(6), 96–108.

Bhambra, G.K. (2021). Colonial global economy: Towards a theoretical reorientation of political economy. *Review of International Political Economy* 28(2), 307–322.

Böröcz, J. (1996). *Leisure migration: A sociological study of tourism*. Pergamon.

Brennan, D. (2007). Love work in a tourist town: Dominican sex workers and resort workers perform at love. In M.B. Padilla, J.S. Hirsch, M. Muñoz-Laboy, R.E. Sember, & R.G. Parker (Eds.). *Love and globalization: Transformation of intimacy in the contemporary world.* Vanderbilt University Press.

Brown, L. (2009). The transformative power of the international sojourn: An ethnographic study of the international student experience. *Annals of Tourism Research* 36(3), 502–521.

Brustein, W.I. (2007). The global campus: Challenges and opportunities for higher education in North America. *Journal of Studies in International Education* 11(3/4), 382–391.

Burtner, J. (2004). *Travel and transgression in the Mundo Maya: Spaces of home and alterity in a Guatemalan tourist market.* PhD diss., University of Texas at Austin.

Cabezas, A.L. (2009). *Economies of desire: Sex and tourism in Cuba and the Dominican Republic.* Temple University Press.

Caton K., & Santos, C.A. (2009). Images of the other: Selling study abroad in a post-colonial world. *Journal of Travel Research* 48(2), 191–204.

Chang, A. (2017). "Call me a little critical if you will": Counterstories of Latinas studying abroad in Guatemala. *Journal of Hispanic Higher Education* 16(1), 3–23.

Conran, M. (2011). They really love me! Intimacy in volunteer tourism. *Annals of Tourism Research* 38(4), 1454–1473.

Cremers, G. (2020). Paradise glossed: Tourism imaginaries, alienation, and the construction of landscape in Santa Cruz la Laguna, Guatemala. *Journal of Tourism and Cultural Change* 18(6), 611–628.

Crossley, E. (2012). Poor but happy: Volunteer tourists' encounters with poverty. *Tourism Geographies: An International Journal of Tourism, Space, Place, and Environment* 14(2), 235–253.

Devine, J. (2014). Counterinsurgency ecotourism in Guatemala's Maya biosphere reserve. *Environment and Planning D: Society and Space* 32(6), 984–1001.

Doerr, N.M. (2013). Do "global citizens" need the parochial cultural other? Discourse of immersion in study abroad and learning-by-doing. *Compare: A Journal of Comparative and International Education* 43(2), 224–243.

Doty, R.L. (1996). *Imperial encounters: The politics of representation in North-South relations*. University of Minnesota Press.

Eberhard-Ruiz, A. (2021). *Jobs diagnostic: Guatemala*. International Bank for Reconstruction and Development/The World Bank. https://openknowledge.worldbank.org/bitstream/handle/10986/35367/Guatemala-Jobs-Diagnostic.pdf?sequence=1&isAllowed=y [August 28, 2021].

Elson, D. (1992). From survival strategies to transformation strategies: Women's needs and structural adjustment. In Beneria, L. & Feldman, S. (Eds.). *Unequal burden: Economic crises, persistent poverty, and women's work*. 26–48. Westview Press.

Elson, D., Grown, C., & Çagatay, N. (2007). Mainstream, heterodox, and feminist trade theory. In Van Staveren, I., Elson, D., Grown, C., & Çagatay, N. (Eds.). *The feminist economics of trade*. 33–52. Routledge.

Enloe. C. (2014). *Bananas, beaches, and bases: Making feminist sense of international politics* (2nd ed.). University of California Press.

Epprecht, M. (2004). Work-study abroad courses in International Development Studies: Some ethical and pedagogical issues. *Canadian Journal of Development Studies* 25(4), 687–706.

Fletcher, R. (2011). Sustaining tourism, sustaining capitalism? The tourism industry's role in global capitalist expansion. *Tourism Geographies: An International Journal of Tourism Space, Place, and Environment* 13(3), 443–461.

Foucalt, M. (1977). *Discipline and punish: The birth of the prison*. Translated from French by Alan Sheridan. Penguin Books.

Grosfoguel, R. (2009). A decolonial approach to political-economy: Transmodernity, border thinking and global coloniality. *Kult 6—Special Issue. Epistemologies of Transformation: The Latin American Decolonial Option and Its Ramifications.*

Guatemalan Ministry for Work & Social Security. (n.d.). *Salarios mínimos fijados por el gobierno de Guatemala para los años comprendidos de 1,955 a 2,021.*

Guatemalan Government. https://www.mintrabajo.gob.gt/images/Historia_de_Salarios_M%C3%ADnimos_Nueva.pdff [accessed August 8, 2021].

Hall, S. (1992). The West and the rest: Discourse and power. In Hall, S. & Gieben, B. (Eds.). *Formations of modernity: Understanding modern societies an introduction.* 275–331. Polity Press.

Hochschild, A.R. (1983). *The managed heart: Commercialization of human feeling.* University of California Press.

Holtgrewe, U. (2007). Restructuring gendered flexibility in organizations: A comparative analysis of call centres in Germany. In Walby, S., Gottfried, H., Gottschall, K. & Osawa, M. (Eds.). *Gendering the knowledge economy: Comparative perspectives.* 250–272. Palgrave Macmillan.

International Association of Language Centres. (2016). *Trends in demand for foreign languages.* International Association of Language Centres. https://www.ialc.org/fileadmin/uploads/ialc/Documents/Study_Travel_Research_Reports/ialc-2016-research-trends-in-demand-for-foreign-languages.pdf [accessed August 28, 2021].

Kempadoo, K. (Ed.). (1999). *Sun, sex, and gold: Tourism and sex work in the Caribbean.* Rowman & Littlefield Publishers.

Kipp, A., Hawkins, R., & Gray, N.J. (2021). Gendered and racialized experiences and subjectivities in volunteer tourism. *Gender, Place, & Culture* 28(1), 45–65.

LaPan, C., Morais, D.B., Wallace, T., Barbieri, C., & Floyd, M.F. (2021). Gender, work, and tourism in the Guatemalan Highlands. *Journal of Sustainable Tourism.* 30(12), 2839–2859.

Little, W.E. (2003). Performing tourism: Maya women's struggles. *Signs: Journal of Women in Culture and Society* 29(2), 527–532.

Little, W.E. (2004). *Mayas in the marketplace: Tourism, globalization, and cultural identity.* University of Texas Press.

Little, W.E. (2005). Getting organized: Political and economic dilemmas for Maya handicraft vendors. *Latin American Perspectives* 32(5), 80–100.

Little, W.E. (2008). A visual political economy of Maya representations in Guatemala, 1931–1944. *Ethnohistory* 55(4), 633–663.

Little, W.E. (2009). Living and selling in the "new violence" of Guatemala. In Little, W.E. & Smith, T.J. (Eds.). *Mayas in postwar Guatemala: Harvest of violence revisited.* University of Alabama.

Little, W.E. (2014). Police and security in the world heritage city of Antigua, Guatemala. *The Journal of Latin American and Caribbean Anthropology* 19(3), 396–417.

Little, W.E. (2019). "Whatever we weave is authentic": Coproducing authenticity in Guatemalan tourism textile markets. In Leite, N.M, Castañeda, Q.E., & Adams, K.M. *The ethnography of tourism: Edward Bruner and beyond.* Lexington Books.

Lozanski, K. (2010). Independent travel: Colonialism, liberalism and the self. *Critical Sociology* 37(4), 465–482.

Lugones, M. (2010). Toward a decolonial feminism. *Hypatia* 25(4), 742–759.

Lyons, K., Hanley, J., Wearing, S., & Neil, J. (2012). Gap year volunteer tourism: Myths of global citizenship. *Annals of Tourism Research* 39(1), 361–378.

Mignolo, W.D. (2002). The geopolitics of knowledge and the colonial difference. *The South Atlantic Quarterly* 101(1), 57–96.
Mignolo, W.D. (2018). Decoloniality and phenomenology: The geopolitics of knowing and epistemic/ontological colonial differences. *The Journal of Speculative Philosophy* 32(3), 360–387.
Mostafanezhad, M. (2013a). The geography of compassion in volunteer tourism. *Tourism Geographies: An International Journal of Tourism Space, Place, and Environment* 15(2), 318–337.
Mostafanezhad, M. (2013b). Getting in touch with your inner Angelina: Celebrity humanitarianism and the cultural politics of gendered generosity in volunteer tourism. *Third World Quarterly* 34(3), 485–499.
Mowforth, M., & Munt, I. (2009). *Tourism and sustainability: Development globalisation and new tourism in the Third World* (3rd ed.). Routledge.
Peterson, J., & Wiens-Tuers, B. (2014). Work time, gender, and inequality: The conundrums of flexibility. *Journal of Economic Issues* 48(2), 387–394.
Quijano, A. (2000). Coloniality of power, Eurocentrism, and Latin America. *Nepantla: Views from South* 1(3), 533–580.
Quiroa, N. (2013). Missionary exegesis of the Popol Vuh: Maya-K'iche' cultural and religious continuity in colonial and contemporary Highland Guatemala. *History of Religions* 53(1), 66–97.
Sanchez, S.M., Scott, K., & Lopez, J.H. (2015). *Guatemala: Closing gaps to generate more inclusive growth*. International Bank for Reconstruction and Development/ The World Bank. [accessed August 28, 2021].
Schneider, M.J. (2018). Exotic place, white space: Racialized volunteer spaces in Honduras. *Sociological Forum* 33(3), 690–711.
Simoni, V. (2014). From tourist to person: The value of intimacy in touristic Cuba. *Journal of Tourism and Cultural Change* 12(3), 280–292.
Simpson, K. (2004). "Doing Devleopment": The gap year, volunteer-tourists and a popular practice of development. *Journal of International Development* 16(5), 681–692.
Simpson, K. (2005). Dropping out or signing up? The professionalisation of youth travel. *Antipode* 37(3), 447–469.
Sinclair, T.M. (Ed.). (1997). *Gender, work and tourism*. Routledge.
Tegelberg, M. (2013). Framing Maya culture: Tourism, representation and the case of Quetzaltenango. *Tourist Studies* 13(1), 81–98.
Urry, J. (2002). *The tourist gaze* (2nd ed.). Sage.
Urry, J., & Larsen, J. (2011). *The tourst gaze 3.0*. Sage.
Van Straveren, I., Elson, D., Grown, C., & Çagatay, N. (Eds.) (2007). *The feminist economics of trade*. Routledge.
Vertovec, S. (2010). Cosmopolitanism. In Knott, K. & McLoughlin, S. (Eds.). *Diasporas: Concepts, intersections, identities*. 63–68. Zed Books.
Vodopivec, B., & Jaffe, R. (2011). Save the world in a week: Volunteer tourism, development, and difference. *European Journal of Development Resarch* 23(1), 111–128.

Vrasti, W. (2013). *Volunteer tourism in the Global South: Giving back in neoliberal times*. Routledge.
Waring, A., & Waring, J. (2009). Looking the part: Embodying the discourse of organizational professionalism in the city. *Current Sociology* 57(3), 344–364.
Witz, A., Warhurst, C., & Nickson, D. (2003). The labor of aesthetics and the aesthetics of organization. *Organization* 10(1), 33–54.
Wolkowitz, C. (2006). *Bodies at work*. Sage.
World Bank. (2017). *Tourism for development: 20 reasons sustainable tourism counts for development*. World Bank Group. http://documents.worldbank.org/curated/en/558121506324624240/pdf/119954-WP-PUBLIC-SustainableTourismDevelopment.pdf.

NOTES

1. I would like to thank Prof. Walter Little for generously reading/commenting on an earlier draft of this chapter. As this chapter derives from my PhD research, I would also like to acknowledge the invaluable support of my supervisors: Prof. John Urry and Prof. Anne-Marie Fortier. All errors are, of course, mine alone.

2. The research was undertaken in 2013 for a PhD on everyday enactments of global tourism competition and, thus, focused on far more than Spanish schools and teachers. The data drawn upon for this chapter includes: one-on-one Spanish lessons with three teachers at two schools; semi-structured interviews with Spanish school directors, owners, and teachers (n=12); and, countless informal conversations with Guatemalans and Westerners. As the interviews were conducted in Spanish, direct quotes are my own translations. Following ethical research practice, I informed "my" teachers that I was a researcher interested in tourism; all interviewees read and signed consent forms; and, all attempts have been made to preserve research participants' anonymity, including using pseudonyms. As there was no space here to discuss what decolonial theory implies for ethnographic research, this will be taken up elsewhere.

3. See INGUAT's Life Lessons promotional videos at https://www.youtube.com/user/VisitGuatemalaVideos/videos.

4. While I use the term Maya here, as many scholars do, it is important to note that many indigenous Guatemalans do not self-identify as Maya, but rather as belonging to one of 20 different "ethnolinguistic" groups (Bell 2012).

5. See INGUAT's "True Wealth Goes beyond Money" video at https://www.youtube.com/watch?v=QZy0ADg3_To.

6. See https://spanishschoolplfm.com/about-us/.

7. See https://news.cision.com/newlink-group/r/guatemala-s-life-lessons-include-learning-how-to-speak-spanish,c9649039.

8. These exchange rates are now historical.

9. This is the minimum monthly salary for agriculture *and* non-agricultural work; the minimum monthly salary for export and maquila work in 2013 was Q1,996.

Chapter 6

Linger: Burned Bambu: Aftermath Nostalgia

Kris Maksymowicz

Linger: (a fragment—the time of ethnography; the time of becoming)

Stay in place longer than necessary because of a reluctance to leave.

—*Oxford English Dictionary 1*

Be slow to disappear or die.

—*Oxford English Dictionary 1.2*

OSKAR AND MARIA

To be slow to disappear or die. This OED definition has begun to haunt me. I think this is because, personally, I have never seriously thought about lingering in relation to actual human death. Crowds linger, slow to disperse after a concert. That person at a party, a quasi-friend, perhaps, lingers, staying longer than necessary—a slowly amplifying or dissipating kind of tension from the scene and the attuned attentions of everyone or no one. Tourists and expats on a beach in Caribbean Costa Rica linger, arriving with imaginaries and intentions of a week in *paradise*, only to be seduced well past their return dates and often their local welcomes. Feelings and sensations linger after scenes of encounter: *I still have butterflies. Lingering doubts and worries.* Or, in more unfinished terms: *Something isn't right. Or, I've got a feeling.*

Ghosts and memories linger, too, haunting the living in a kind of refusal to die. Have-beens linger beyond their expiration dates in the form of out-of-touch politicians or washed-up actors, athletes, or musicians turned reality television stars. *Oh, why the fuck can't Donald Trump just die, already!* I heard someone say the other day. But, of course, they don't really wish that Donald Trump was dead . . . probably. I wish you were dead (or I was never born), yelled almost every teenager in the hyperbolic pitch of an argument.

But. To be slow to disappear or die. Like, really. *Slow Death* (If you're a fan of Lauren Berlant 2011). I never thought I would encounter something like this in a way that wasn't figurative or facetious. Then came Oskar and Maria and a beach party barbeque scene on the edges of neoliberal empire on the Caribbean coast of Costa Rica (see Little 2014). Like, really, *man.* Crack-addled, scuttling, hustling, fucking, drunken, Oskar and Maria, lingering; being slow to disappear or die. Or, is this some kind of stubborn refusal? Making life-worlds in dire straights. How does this come to matter?

Picture this. Jake just built a barbeque. In the wake of a Limón gangster fiasco, Jake bought a bar following a Scarface style showdown where the previous owners, Cia and Thomas, had to sneak out an escape hatch while pistols and automatic weapons trained their sites on the front door of their upstairs apartment. Months back, Thomas ran a poker night that got held up when the betting was getting heavy. Word is some thieves came in, pointed weapons, and ran off with a hefty chunk of change.

Everyone was rattled.

The local expat community banded together. Things were getting scary in Puerto Viejo and enough was enough. *We'll back you*, they all said. *We need to stand up to these criminals; we need to take back our town*, others reprised. An echo chamber of expat concern, worry, agitations, nervousness, and un-reflexive ownership in Atlantic Costa Rica. Private expat Facebook forums became supercharged with strategizations. People talked tough. Others talked tough through online handles. There was paranoia. *You don't really believe this forum is private, do you?* Everyone knew who the *thugs* were. The public secret was out.

Cia and Thomas filed a *denuncia*[1] with the *Organismo de Investigacion Judicial*[2] in *Bribri.*[3] Things didn't go as planned. *Drug dealers* from Limón started showing up in their bar, telling them to *back the fuck off*, or else. *Don't break ranks. Hold fast. Keep in line. Call their bluff.* But, of course, there was no bluff. After the warnings came some serious intimidation. A line of armed gunmen organized themselves in front of the bar one night following the rumors of police involvement. Cia and Thomas snuck out the back. *It was like the fucking Alamo*, Mark tells me. *They had to go into witness protection in San Jose*, echoed Annie. Others tell me that that's not how it went down at all. *Why would people in witness protection post pictures of their house in San Jose?* Confusion-gossip-contradiction-hyperbole-worry-pile-up.

In the days and weeks that followed, those carefully tightened expat ranks began to fracture and lines were drawn, gesturing different modalities of discipline in the paradisiacal town. Some people fell off the map, fearing reprisal on themselves, their businesses, or their families. Other expats became reactionary, spitting contempt for those sensationalizing the event, Cia and Thomas for flaunting that kind of cash in a place like Puerto (*what did they think was going to happen?*), and stifling claims about violent trouble in paradise. Others, still, began chastising "deniers," championing the need for transparent discussion about the crime wave cresting in their beachfront slice of heaven (*denial is only going to make things worse*).

The United States issued a travel warning. Business did get worse. Sketches of neighborhood watch programs were drafted and meetings were organized—location TBD for those *right kind* of expats "on the need to know" that needed to know. Donations were collected for a series of CCTV cameras that never ended up working. A beach security team was implemented. The US Embassy gave accolades to an expat watch-group for its exemplary foreign lead crime reduction initiatives. People made graphs. Some people left for one bifurcated reason or another. *This place is getting too crazy* or *I didn't come to Costa Rica to live in a police state.* Layer upon layer of amassing complications.

Cia and Thomas left, too. They had *had enough*, they said. The dream life was ruined. They put their bar up for sale and moved to San Jose while lawyers handled the business. Jake bought the bar. Then Jake built a barbeque. *Sunday Funday* was born.

The whole thing started off slowly. Jake just liked burgers and barbequing. If you bought a beer the first burger was free, any more cost 1500crc (about three dollars). Jake sourced good beef (*Have you seen the cows in Costa Rica? Abysmal*), found a place to buy imported American Cheddar (*Real American cheese? I can't believe it*), and started serving the biggest Bloody Mary's the Caribbean had to offer (*Good God, is that candied bacon in there?*). People got wind. Before anyone knew it, scores of expats were headed down to Jake's on Sunday nights to slug burgers and down beer in cheap patio furniture, set to the backdrop of a black sand beach and a fiery mango sunset.

It was then that things started to get complicated, again. At first, the issues were minor. Jake kept running out of clean dishes and glasses. Disposable plates, cups, and cutlery—problem solved. Next, garbage started piling up on tables. Ketchup-mustard-mayo-beer-sweat-swirls getting on peoples" elbows and clothes. Like many bars in town, Jake hired "*crack-heads*" to bus tables in exchange for a drink and a meal he was already making. More locals started showing up for the cheap food, to hustle bleeding-heart expats and longs stays, to get laid, or to party. Next thing anyone knew, the *Sunday-Funday-Expat-Secret* became a site of agitation. There was lamentation. There was complaining. There was tough talk. Expats started getting mad over little things like how long they had to wait for a burger or that they couldn't leave their cigarettes on the table (*these fucking crack-heads have eyesight like Superman*). Things started to feel like a pressure cooker. Soon enough, the drunks and druggies that lived under the bridge next to *crack alley* were stumbling and fidgeting their way toward that *mango sunset* that looked like *fyah*, working to hustle a quick buck for a hit or a drink. Maria bused tables. Oskar hustled.

Oskar and Maria worked as a team. Unlikely bedfellows (Maria, an Afro-Caribbean woman in her fifties, and Oskar, a young man of mixed Caribbean and Indigenous descent in his early twenties), what the pair lacked in an aged and raced social cohesion, they made up for in shared, out-of-control addictions and stories of abandonment. Recently released from prison for what was rumored to be an incident that escalated from break-up, to crack bender, to break-and-enter, to attempted stabbing, Oskar was on parole and off the wagon. In his moments of clarity, Oskar was a voracious storyteller and a subtle critic, lamenting an addiction that had led to quarrelsome and silent relationships with most of his family and children, and venting thinly veiled contempt for a disinterested government and an aggressive tourism industry. At his worst, Oskar was a veritable wrecking-ball of affective force, stumbling and screaming, making threats, wielding knives, listening and singing to music that wasn't playing from unplugged headphones, and conjuring anything from a *tourism-party-excess* that caught *go-hard* vacationers up in his inertia, to disgust, derision, rage, avoidance, pity, or who knows what else for those who *were sick of his shit*.

Maria was like Oskar plus thirty years. Coronated by the expat community as *Queen of the Crack Heads*, Maria held a kind of institutionalized place in Puerto Viejo that ricocheted between reverence, mockery, and fear. During the day, yoga teachers kissed her on the cheek and gave her hugs in front of the Market, patting her on the back while asking how she was doing. Behind that patted back, people told stories about her antics in ways that always morphed into some kind of stand-in for how *crazy* life in the town was becoming. *Do you remember the time Maria . . . This town is crazy*. For a while, people liked to circulate security camera footage of Maria fucked up or cleaning the shit some wrecked tourist accidentally let loose on an unsuspecting dance floor as part of her busing gig, laughing hysterically at a monstrous, fuzzy, black and white scene that scattered vacationers in full on party mode. *Look at her go!*

People also warned about her unpredictability and violence. *Don't turn your back to her; she won't hesitate to stab you.*

After Oskar had been kicked off of his family's farm, he took up a tenuous residence in a shanty on a small plot of land that Maria had acquired by some means. When he could sleep, he slept there, in a hammock that was constantly wet from the rain. It was in that shack that Maria and Oskar took care of each other, cooking food that they had cobbled together throughout the day, cleaning infections on their bare feet, and sharing drugs if they had any to share. When Maria hadn't seen Oskar for a few days, she asked after him, moving from bar to bar, party to party, hoping to find him, to see if anyone had seen him, and telling anyone who would seriously listen that she was worried about him. Despite their erratic schedules, their disappearances, and their addictions that would lead them all over town, it was certain that, on *Sunday Funday*, they would see each other and work the crowd of vacationers and expats who had gathered for the evening.

Around five o'clock, Jake fired up the barbeque. By five thirty, every wooden table on the beach in front of the bar was full, lined with expats both salivating and eager for burgers and Bloody Mary's. While waiting, people walked on the black sand beach into calm, lapping, azure water, taking pictures of an "impossibly" (see Little 2014) fiery and vibrant sunset pulling behind the *Talamanca Mountains* and uploading them to social media with cliché hashtags like "*#wish you were here*" or "*#another day in paradise*." The whole ritual was a weekly, continuously unfolding sensual exercise in nurturing a kind of disbelief, an act of repetition building into a habit of *counting your blessings* and not being able to *believe the beauty* of yet another perfect sunset in the Caribbean. Chilled out Bob Marley tracks played in the background. The twangs and drawls of a hundred different accents, the sizzling of barbeque, and the push-pull sound of a calm Atlantic Ocean textured the evening as the sun went down. The smell of ganja seemed omnipresent. *This is bliss*, people repeated aloud in a pseudo-local-Caribbean parlance, like some kind of mantra, whether they felt that way or not. *This is why we made a crack at life down here,* they would explain. *This is why we hang around . . . in spite of the bullshit . . . Pura Vida.*

Pura Vida. Pure Life. Monstrously complex. The Costa Rican motto is a slippery cultural pillar in the country's *everyday*; a deceivingly difficult, often performatively glib kind of statement that operates as anything from a greeting, to a "shit happens" circumstantial acceptance, to a stripped-down "natural" conceptualization of life, to an entire national/cultural ethos of taking things as they come. Recently, however, *Pura Vida* is also something that is being multiplied and continually (re)made and folded into Costa Rica's emergent tourism markets and scenes. Commodity garbage, the kind of stuff you can buy in airport gift shops and hotel lobbies is scrawled with the phrase, alongside depictions of *perfect* beaches and jungles, adding to touristically in-filled "tropicalizations" (see Thompson 2006) and naturalizations of the country and some of its people. Government funded tourism advertisements and beer commercials stitch *Pura Vida* into their Marketing campaigns (see the esto es Pura Vida Imperial Beer commercial, 2014), hawking a laid-back natural wonderland that is safe and inviting for locals and tourists alike (Rivers-Moore 2007).

Scale down. Fast forward. Tourists bought in. Locals did, too—continually refashioning themselves in the wake of emergent tourism demands and trends (see Anderson 2003; Frohlick 2013). Everything changing through imaginaries and the encounter dance (Faier and Rofel 2014; Bruner 2004; Pratt 1992; Little 2014). Subjectivities, people, bodies, spaces, and desires unfolding in the contactive scenes of Puerto Viejo's emergent neoliberal tourism economy. And before anyone knew it, Pura Vida had become a complex cultural cornerstone in the touristic everyday of Puerto Viejo; a fidgety, positioned, ricocheting, shifty assemblage cobbled together from circulating local and neoliberal imaginaries, and defined and undefined sensory overloads.

Question (for expats). *What does Pura Vida mean to you?* Answer. *Oh, you know, it's just life in its purest form. Pure Life. Hard to explain. It's the sounds of the jungle. The taste of fresh fruit. Living with less. Toucans and sloths. Hakuna Matata. Finding love with locals in the tropics. Sand between your toes. Not having shoes.*

Tourists were hooked. Some people got Pura Vida tattoos. Some moved halfway across the world just to get a little more Pura Vida. Locals started hawking, in one way or another, experiences entwined with such material-semiotic imaginaries, expectations, longings, and desires. Awkward attachments were formed. Appropriations were made. A kilometer into the dense rainforest, a foreign owned yoga-nature resort sprang up, selling *sustainable* lifestyle and consciousness raising experiences "connected to the culture of Pura Vida," and tailored to "citizens of the world" (see Samasati Nature Resort promotional videos). Online, *This American Girl* began peddling courses and retreats on *"authenticity"* to an emergent travel *"tribe"* from lessons learned through living Pura Vida on the Atlantic coast of Costa Rica. Expats, long-stays, tourists, and locals alike started getting blackout drunk in the morning. *Fuck it! Pura Vida, Right? Find your bliss.*

So there we were, an anthropologist and some expats, some yogis and some tourists, sitting around wooden tables on a black sand beach while a mango sunset sank over the Talamanca Mountains, each living some kind of touristically in-filled Pura Vida. Each of us differently lingering, by way of different histories and relationships with uneven global tourism economies, in the contactive, effervescent space of Puerto Viejo as an emergent encounter zone (Bruner 2004). And that's when it happened. As people got their fill, got drunk, or stoned, or both, Maria started her bussing gig on a hot Sunday night in May of 2016. In a slinky summer dress and a red, green, and gold bandana, Maria slid around tables and begging dogs, cleaning up garbage and wiping up messes, hustling a quick tip where and when she could. For the most part, people ignored her, sunk into this or that conversation about local gossip and happenings. But, as she made her way to one table in particular, somebody said something, and Maria lost it.

Four words. "*Gracias, Maria. Pura Vida.*" Everyone was expecting a "*de nada.*" But with that, Maria went off like a shotgun. In her booming, raspy voice, Maria screamed: "*Don't talk that fucking shit to me. Talk to me in my language. English. I don't want none of this Pura Vida shit. Fucking tourists. This place isn't called Puerto Viejo. It's Old Harbour. Wolaba. Fuck you. Go home.*"

From the nearby beach, Oskar rushed over. With a few words about needing the job and the money, he calmed Maria and worked to smooth the nervous and agitated edge of the roughed-up scene. Heart rates started to fall. Hands stopped shaking. Having Maria's back, Oskar asked that nobody make a big deal or say anything. *Everyone cool?* Everyone agreed. With that, Oskar bummed a few cigarettes, a couple of *snappers*[4] for drinks, and took his leave, off to move through the crowd, looking for people he might know or be able to hustle.

But Oskar's smoothing work wasn't totalizing. He couldn't fully contain the force of Maria's outburst or her presence. In that moment of contactive encounter, Maria instantiated something, shifting the chilled-out atmospherics (Stewart 2011) of burger night, cliché photography, Pura Vida, and a happy-go-lucky, expat buzz in a paradisiacal town at the outskirts and frontiers (see Frohlick 2013) of Costa Rican tourism circuits.

"*I'm glad that's over,*" Martin started. "*I mean, I don't like to have my back to Maria. You never know what she's going to do. Did you know she . . . ? It was like some kind of. . . . Do you remember the time Maria . . . ? Man, I can't believe she's not dead yet.*"

What Maria set off was a deeply troubling series of recollections and storytelling about *how crazy Puerto had become* in the wake of drug and tourism infiltration. Stories about rampant addiction and crazy happenings that you *couldn't make up*, and a hyperbolic disbelief, that, when uttered, didn't feel so hyperbolic. About the kind of *element* that was roughing up the place. About how, *this element, "these people,"* were still *lingering*. About how people at the table *couldn't believe how long it was taking "those people" to die or that they were still alive, at all*, and about how expats living there stick around for the sunset, *in spite of the bullshit.* Pura Vida.

Temptation

At this moment, I feel tempted. Specifically, I feel tempted to mobilize the above entanglement of stories as something one might call an "ethnographic vignette." In ethnography, the vignette has served, I think, two very important functions: the first, as one of many strategies to breathe a kind of liveliness into a practice that, I argue, without, would be quite dull and removed; and second, as a complex true-fiction (see Geertz 1973; Fabian 1983; Clifford and Marcus 1986; Marcus and Fischer 1986) from which, and through which, ethnographers studying tourism make, render, and qualify some kind of positioned, partial, descriptive, and analytical point about what they are studying—usually an argument about the complicated, historicized, and power laden dance of encounters, wherein novel ideological apparatuses, embodiments, imaginaries, interpolations, and subjectivity formations relationally, yet differentially, unfold through tourism contact (see Faier and Rofel, 2014). The same could be said, I think, about ethnography as a practice, more generally. Let's call this representation.

And why shouldn't I be tempted? Think about it. Imagine this chapter working to take up the stuff of these stories and rendering it into some kind of counter-narrative about tourism emergence in Atlantic Costa Rica, in opposition to that same old story about how, through tourism development, money *was supposed to start pouring from the sky and about how everyone was supposed to get a piece?* (Little Forthcoming) About how the sordid and cunning (Povinelli 2011) history of a racist state has worked to marginalize and "other" Afro-Caribbean and Indigenous bodies in Costa Rica through a framework of racialized homogeneity based on "white" colonial Spanish descent that obfuscates the Limon Province within Costa Rican national identity, hawked for the purposes of fashioning Costa Rica as a safe and welcoming tourism destination for all (Palmer and Molina 2004; Anderson 2004; Rivers-Moore 2007). About how tourism advertising campaigns have worked to "naturalize" and complexly (and transgressively) "heterosexualize" and "eroticize" black bodies in Puerto Viejo, the Limon Province, and the "Caribbean," more generally (Rivers-Moore 2007; Frohlick 2013; Frohlick and Johnston 2011). And about how drug addicted and other touristically non-conforming local black and Indigenous bodies are being subjectively (re) fashioned through power-laden tourism contact as problematically *lingering* in a paradise at the end of the road—that is, as being too slow to disappear or die, or, sometimes, as the stuff of *bullshit.*

Because in her outburst, that's pretty much what Maria did. In five seconds, Maria deterritorialized the official and seductive imagery of a tourism impulse-bliss, unleashing an unwavering and aggressive critique of nation-state politics and touristic processes, manifested through everyday things like Spanish language imposition, through not so simple things like what a place is called and by whom, and about how many tourists, who are in the thick of it, and roped into it, have no idea what's going on or what they're contributing to. This is to say that, almost instantaneously, Maria challenged the reforming politics of a tourism-imaginary-machine with powerful recalibrating capabilities, specifically, those with the capacities to (re)fashion *Pura Vida* and the bodies of local people like herself and Oskar as *trouble*, *troubled*, and as the stuff of *bullshit*.

And through the confrontation, through the affective force of her verbal tirade, Maria instantiated a rupture that broke the smooth and predictable unfolding of a beach party barbeque scene in a tropical tourism paradise. In that moment, Maria bravely worked to violently rip open a wound that had been festering beneath a sutured surface, potentializing an everyday event that was on track to play out how it should, until it didn't.

Unless, it did.
Recalibration is cruel.
Power is cruel.
Habit is cruel.
Crisis ordinaries are cruel.

What I mean is, in a place like Puerto Viejo, where the ordinary can often feel like a crisis, potentialized irruptions and instances such as Maria's "calling out" aren't exactly all that uncommon—but neither are their attempted resuturings. To be perfectly honest, this wasn't the first time I had witnessed expats and long-stays refer to "troubled" and "troubling" local people like Oskar and Maria as unsavory *elements*, or as being too slow to disappear (by way of an imminent, yet too slow death, or otherwise) from their *Pura Vida-Paradise-Bliss*, as if through some form of habit "residing in the flesh" (Massumi 2002: 11), what Massumi (1992) might call the "ballast that chains a dog to its vomit" (47). Either way, to me, the question seems moot. It all adds up to circuits of rupture-reterritorialization. But what of Maria's affective excess? Because, when I asked her about it later, Maria couldn't remember why she had said those things.

In the days and weeks that followed, this scene of touristic encounter began to weigh on me. How could Maria's potentialized "outburst," what had arguably amounted to, or could be ethnographically rendered as a shocking and astute critique of historical, national, and touristic emergence that had short-circuited and derailed the trajectory of a happy-go-lucky barbeque in *paradise*, been disregarded, and worse, used as a catalyst to launch its witnesses into an habitualized wave of recollections and storytelling that effectively worked to (re)- and continually produce her already multiply interpolated body as problematically *lingering* and the stuff of *bullshit*? And what else might an anthropology of tourism do with such affectively potentialized moments of shocking irruption? And what of Maria's inability to remember her speech act?

Betrayal

Don't talk that fucking shit to me. Fuck you. Go home.
Remember the *time* Maria . . .
I can't remember.
Maria Betrayed; Lingering as Encompassment; Time as Measure.
Lingering Betrayed; Time as Duration, or, of a different order.
I can't believe Maria is still alive.

Here's the thing about *lingering*, as a differentially valued and interpreted kind of subjectivity encompassment enacted at the convergence of already multiply historicized and interpolated bodies. Its very mention betrays its politicized encompassing trajectories. Unlike material-semiotic assemblages (Deleuze and Guattari 1988; Massumi 2002) such as race, gender, and sexuality, etc. (see Saldanha 2007), wherein aggressive processes of historicized naturalization and production have obfuscated their relational ongoingness (gender is, for example, after all, always a relational "doing," a gender-ing, if you will) lingering flaunts its unfinished unfolding, hiding in plain sight. In Puerto Viejo, multiple and complexly interpolated bodies in contact come to be understood, felt, and encountered as differentially lingering, for better or for worse, through touristic circuits of encounter-subjectification/differentiation. It is the enacted politics of this lingering that comes to make up the stuff of *bullshit bodies* and *too slow death*, or the punch lines of not-so-lighthearted jokes about tourists and long stays pushing back plane tickets to sit on a beach in *paradise,* or about being sucked indefinitely into what I have elsewhere conceptualized as the *party vortex* (Author: 2018).

But, here's the rub. Staying in place longer than necessary because of reluctance to leave isn't gone nor settled. And the horrifying abjection of slow death and maintenance that comprises the everyday for "non-conforming" local bodies in contactive tourism encounters isn't yet over, crushing and confining though they may be. Where there is movement, there is possibility, however modest.

My goal, then, here, is to let the unfinished nature of lingering, as an ongoing, material-semiotic encompassment with differential and differentiating capacities for relational affectation, betray itself through a politics of mimetic extension. My goal is to amplify movement toward possibility; to kick a habit, here understood as an "acquired automatic self-regulation";[5] to stubbornly stick with the stubbornness; to stay in the middle; or to see where things might go if left unchecked. But what of the quality of this movement? And what of its temporalities? Let me explain.

While I am seductively tempted to take up the stuff of Maria's outburst and, once more, render it into some kind of vignette that works toward interrogating the politics of her produced subjectivity positioning as an intersectionally contained and "problematically" lingering body, I want to begin to ask different kinds of questions about such moments of irruption. Specifically, rather than folding such deterritorializing moments into an ethnographic vignette under what Ochoa (2007: 49) has called "the three pillars of anthropological representation and insight: negation, identity, and being," I want to begin to ask questions about the affective and rupturing force of events like Maria's outburst, before they become habitually and powerfully reterritorialized in backgridded memories and narratives—locally, ethnographically, representationally, or otherwise. Put simply, I want to think seriously about how Maria evades (not simply resists) her (re)subjectification as a "troubled lingering body." I want to think about how Maria and Oskar continue to endure and modestly build worlds, even under the crushing and confining, traumatic and unequal auspices of global tourism practices in Puerto Viejo and Costa Rica.

To this effect, I seek to rearticulate a notion of *lingering*; to understand it not simply as a deeply problematic kind of subjectification produced in the uneven, yet liminally creative space of the touristic border- or encounter-zone (Bruner 2005), but as an *event*. As Massumi (1992), following Deleuze and Guattari (1988), notes, an event is "neither stable nor transcendental nor eternal; it is immanent to the dynamic processes it expresses and has only an abyssal present infinitely fractured into past and future" (Massumi 1992: 18). Contrary to the stylistic "displacements" of representational theory, wherein a politically charged "ethnographic present" (Fabian 1983; Clifford and Marcus 1986; Marcus and Fischer 1986) renders bodies out of time, as if they were "leaping from one definition to the next" (Massumi 2002: 3), understanding lingering as an event means nurturing and extending a kind of betrayal. It means understanding that the "unity of the essence [of a form] is always self-undermining" (Massumi 1992: 18) and in motion, continually betraying its own stability.

For cultural theorists like Massumi (1992, 2002), specifically, those who are working within the scope of the "non-representational turn" (Gregg and Seigeworth 2010; Thrift 2008; Manning 2016), *events*, what in representational ethnographic parlance might be characterized as anything from subjectivities, to objects, to publics, to epochs, etc., figure as more-than-human, assembled, and relational stabilizations and subtractions of affective capacities to affect and be affected, although never fully. Here, affects operate as "embodied and impersonal . . . autonomous domains of intensity, indeterminacy, and above all, *potentiality*" (Mazzarella 2009: 292 emphasis mine; see also Massumi 2002). In this vein, affects are neither intentional nor unintentional: "they do not mediate the body through a signifying image . . . they are immediate and they are always co-present, not in a body but 'of bodies' in a relational transition of entanglement, or bodies coming together as an open-ended assemblage of entangled ways of living that cross-up multiple temporal rhythms for better or for worse" (Little: Forthcoming). This is to say that affects are of a different order—not a-cultural, but pre-cultural—wherein their potentialized intensities and indeterminacies are "backgridded" (see Massumi 2002) and foreclosed through the encroachment of powerful signifying logics "after the fact," what Massumi (2002) has dubbed "the half second delay." For critics like Manning (2016) and Massumi (2002), however, there is always an escape of affect, the process of which engenders, not only, the relational emergence and unfolding of novel bodies and worlds, but also, a paradox for representational ethnographic accounts of identification and subjectification, heretofore.

In relation to objectification, for example, Manning (2016) writes: "[There] is a paradox: for there to be a theory of the 'object,' the 'object' has to be conceived as out of time, relegated beyond experience, unchanging" (48). Drawing on Whitehead and an example of a chair, Manning argues: "It is only after the fact, after the initial entrainment the chair activates, after the movement into the relational field of 'sitability,' that the chair as such is ascertained, felt in all its 'object-like' intensity" (48). Manning continues:

> The object *is* the abeyance—the feeling-form (a form felt more than actualized) that cannot be separated out from the milieu, from the field that it co-activates: in this case something like the ecology of comfort, sitability, and desire to sit . . . The object, like the subject, is never it-self. (48)

In her recent book, *The Minor Gesture* (2016), from which these quotations are drawn, Manning argues for the development of a cultural theory that more convincingly accounts for how experience is "actually experienced," that is, as a relational "field of expression through which different quality[ies] of experience [are] crafted" (47). For Manning, the qualification of an event (which may amount to the production of a "body," a "subjectivity," an "object," etc., or a lingering iteration of any of these), requires a *measurement* of the time of the event through the activity of backgridding affective intensity and the encroachment of signifying logics. Such an act, as it occurs "after the fact," "tends to deplete the event-time of its middling, deactivating the relational movement that was precisely the event-time's force" (47). Here, anything out if it's middling, that is, "out of movement, is out of act . . . [and]

[a]nything that stands still—an object, a form, a being—is an abstraction (in the most commonsense notion of the term) from experience" (47). So why does any of this matter?

While deeply complex and multifaceted, Manning's (2016) overarching argument is relatively digestible. It goes something like this. That for complex reasons (which I address later in this chapter), and although usually politically oppositional to the effects of powerful signifying logics that create what anthropologists have come to understand as things like "subjectivities," academics have come to be complicit in replicating the practices of powerful structural tendencies, spilling ink critiquing, debunking, and deconstructing abstractions; out of time, out of movement, and out of act. While not necessarily problematic on their own, such tendencies, as Stewart McLean notes, risk "enshrining a normative empiricism that absolutizes existing actualities as their unchallengeable horizon of what might 'count' as reality . . . [producing] a too ready acceptance not only of hegemonic criteria of 'interest' and 'relevance,' but also a kind of ontological status quo" (in Romero 2015).

To clarify, affective excess—those intensities which "exceed the bounds of the event, touching on the ineffable quality of its more-than" (Manning 2016), those indeterminacies, couched in what Manning (2016), after Deleuze and Guattari (1988), calls "minor gestures"; those singularly connected and inventive potentialities that murmur in the relational interstices of structural tendencies and that activate new worlds, modes of being, and other ways of living—is, in the current academic climate, being "narrated as secondary, or even negligible" (Manning 2016: 1), or altogether "abbreviated from [anthropological] sensibilities of the world" (Shapiro 2015: 372).

For Manning (2016), and those involved in such conversations (Massumi 2002; Romero 2015; Mclean 2015; Shapiro 2015; Little 2014), what is ultimately at stake is the perpetuation of an account of experience (and ultimately, a mode of practicing "ethnography') that neglects affective political affirmation, that is, that forgets "the movement activated, in the event, by a difference in register that awakens new modes of encounter and creates new forms of life-living" (8).

To take up lingering, solely, as an abstracted[6] encompassment or type of subjectification, wherein it becomes possible to tease out a critical analysis of which bodies differentially linger, and in what ways, ultimately (and rightly so), toward the debunking of racist valuations that are, literally, commentaries on who deserves to live or die, when, how quickly, and under what conditions in a Costa Rican tourist town, is also, complexly, to participate in a conjuring trick that obfuscates the fact that to linger is to be unfinished and in motion. It is to render lingering out of motion; out of act. This is to say that it is to *measure lingering's event-time*. Put simply, to suggest that Maria or Oskar "linger," or "linger improperly," or conversely, to be solely critical of the politics enshrined in a complicated disciplinary act that encourages them to "hurry-up and die," is ironically, to strip lingering, something that is decidedly unfinished, of its durative temporal force.

Let me be perfectly clear on this point. What I am not doing is suggesting, in any way, that critical assessment or commentary is somehow pointless or passé, something akin to the caricature critiques leveled in favor or against "the affective turn," wherein absurd polarizations render affect studies as either the privileged way out of representational impasses, or as a flighty digression that "requires getting back to the real stuff" (see Hemmings 2005: 548). It matters, and it matters immensely, and it forms the basis for affirmative augmentation (see Massumi 2002). What I am suggesting, instead, is that there is more to the story; that there is always more there (Thrift 2008). What I am grasping for, is a way to deal with things like Maria's inability to remember her speech act; those things that "don't lend themselves to a perfect, three-tiered parallelism between analytic subject, concept, and world" (Stewart 2007: 4); those things that are the stuff of vital emergence that literally press bodies toward new infrastructures of difference and sensation.

In taking up *lingering-as-event*, I seek to reinstate a type of middling (see Manning 2016) to the ethnographic concept. This is not a question, "once again, of slow time, or quick time, of lingering [as measure or valuation– my inclusion] or speeding" (Manning 2016: 52–53). Event-time is not quantifiably measurable. Instead, "[i]t is a question of moving experience beyond the way it has a habit of taking, of discovering how the edges of life-living comingle with the force of that which cannot yet be perceived, but nonetheless felt" (Manning 2016: 53). It is about finding some wiggle-room in the interstices of Maria's outburst, and, in ethnography, it is about committing an act of what Kathleen Stewart might call *unforgetting* (1996:71).

Unforgetting

> *Or, "what would happen if I tried to unforget"?* (Stewart 1996: 71).

I don't remember, Maria said to me when I asked her about the outburst. *I don't remember saying those things.* She was hustling me for a *pachita*[7] of *guaro*[8] and a bag of onions when I posed the question. *How could you not remember?* I pressed in passing. *I don't remember a lot of things,* is all she said.

This wasn't some wink pulled from a cinematic cliché, where a bartender suddenly remembers more when a twenty-dollar bill slides across a counter. Maria either didn't remember or she wasn't going to tell me. And I had already bought her the booze.

The moment tasted like ethnographic defeat. A lost opportunity to write a story of back-talk and modest resistance directed toward aggressive global and state-sponsored tourism practices and the people who took stake in their propagations, wittingly or unwittingly, to the detriment of local bodies that ended up shrapnel in their deterministic and signifying wake. A lost opportunity to write a different kind of story about tourism emergence in a Costa Rican tourism town; to get the ethnographic dirt on how locals push back against their subjectification, against all odds; about how people are never totally subsumed by their circumstances. Maybe it still is that. And it is still possible to tell Maria's story without those details. After all, it would be easy enough to forget that she forgot.

But what of Stewart McLean's (in Romero 2015), Erin Manning's (2016), and Nicholas Shapiro's (2015) provocative warnings about anthropological abbreviations of those sensibilities that exceed scenes of encounter; those things that "don't lend themselves to a perfect, three-tiered parallelism between analytic subject, concept, and world" (Stewart 2007: 4); those things that are "instead, a problem or question emergent in disparate scenes and incommensurate forms and registers" (Stewart 2007: 4)?

Because what was so intriguing about Maria's outburst was that it wouldn't leave me alone.

What was happening?

Don't worry about it, I thought. *Just forget about it.*

But the question gnawed on me as I lay in bed at night, returning, again and again, in refrained waves of obsession. A flicker becoming fever pitch *intensifying* in the riotous and frenzied chaos of her tirade's unlocatability. A shocking, flash-bang speech-act overwrought with unspeakables, demanding more of my anthropological sensibilities and literal ways of making sense.

(Refrain)
Affect is intensity.
Intensity that is not amnesia.
Not remembering that is not amnesia.
The body registers.

It would be easy enough to write that vignette. It would be easy enough to abbreviate that form from my ethnography—to simply funnel Maria's outburst into some kind of representation of lingering as yet another layer added on to an already long and multiply interpolated colonial-touristic process of subjectification—even though I could feel the "absence" in her memory working on me. But that's the point, isn't it?

That I could feel it.

Working on me.

This is a question of becoming sensate. This is a question of committing an act of *ethnographic unforgetting*.

(Refrain)
Affect is intensity.
Intensity that is not amnesia.
Not remembering that is not amnesia.
The body registers.

Affects are neither intentional nor unintentional. They "don't mediate the body through a signifying body image, they are immediate and they are always co-present, not in a body but 'of bodies' in a relational transition of entanglement, or bodies coming together as an open-ended assemblage of entangled ways of living" (Little: Forthcoming). This is another way of saying that Maria's "forgetting" *matters*. Literally.

What is so important about Maria's inability to remember, whether purposeful or not, is the fact that it traffics purely in intensity. As a "ghostly mattering" emanation (see Gordon 1997), Maria's verbal performance and her "not remembering" haunts any "finished concept," standing "in excess of any cultural 'system' I [or anyone else] might have conjured up" (Stewart 1996: 71).

"What would happen if I tried to unforget?"
(Refrain)

Affect is intensity.
Intensity that is not amnesia.
Not remembering that is not amnesia.
The body registers.

In 1996, through unforgetting, Kathleen Stewart was carefully critiquing modes of ethnographic authority in the wake of the *Writing Culture* (Clifford and Marcus 1986) movement. In doing so, Stewart took pointed concern with anthropological treatments of the anecdotal, tracing, after Clifford (1986), a long trajectory of mobilizing the anecdote toward the establishment of ethnographic authority, particularly in the contexts of the uniqueness of "having been there," and in terms of overcoming ethnographic naiveté in the early stages of fieldwork, a process that has come to be couched in what has been termed the "arrival narrative."

Following Mary Louise Pratt (1986), Stewart (1996) astutely draws attention to the hierarchization of the "scientific, self effacing" (74) characteristics of "ethnography proper" over the supplemental status of the anecdotal, rendered most forcefully in the bifurcation of anthropological publications, wherein "secondary autobiographical text[s] that [spoke] to the experience of fieldwork" (74) became a common, yet marginal, feature of the discipline. As argued by Stewart (1996), such practices closely mimic and "ally with a dualism of 'subjective' and 'objective' knowledge, [which], as distinct texts . . . belie a hierarchical order of accounts" (74) and, she suggests, encase *events* "in an endless replay of a desired othering that is contained and held at a distance—the closed loop that Said (1978) called Orientalism, that Bhabba (1990) called colonialist ambivalence, or that Stocking (1989) called romantic motives" (72).

Arguably, the rigid bifurcations Stewart was describing in *A Space on the Side of the Road* (1996) have become less pronounced in recent years, with ethnographic writing conventions propagated by the *Writing Culture* movement having taken serious hold in the genre that is the corner stone of the discipline. To this effect, anthropology has witnessed a surge in the production of dialogical (see Stewart 1996), historicized (Comaroff and Comaroff 1992), surrealist (e.g., Taussig 2004), anecdotal, and reflexive ethnographic publications—specifically, those that have made significant efforts to decolonize the discipline, as well as, highlight the relationality of knowledge production in anthropology.

And yet, in spite of efforts to break anthropology from problematic and deeply Eurocentric habitual renderings of alterity as "out of time" and "over there" (see Fabian 1983), a process ultimately geared toward collapsing the subject/object binaries that undergird many of the theoretical foundations of the discipline, anthropologists have come to be consumed with refining or illuminating the limits of ethnographic conventions toward conversations regarding the positioned and partial nature of that which they describe in a deeply presentist mode of ethnography, interested almost exclusively in *what* gets produced in the contactive and creative space of encounters across difference, and the politics thereof. In this sense, the anecdote or vignette, has come to be mobilized as a tool in the layering and concretization of anthropological truth claims, albeit partial ones.

But positioned, partial, and presentist ethnography is different from that which insists on reinstating a type of middling to the genre. Indeed, it is precisely the assertion of Manning (2016), articulated earlier in this chapter, that in order to theoretically account for the *what* of cultural production (a "body," a "space," an "object," a "subjectivity," etc.), theories of *things* necessarily constitute them as "beyond experience . . . out of motion, out of act" (48); carrying echoes and traces of the very distinctions, binaries, and temporal distortions such efforts were designed to circumvent. Is it any wonder, then, that I am tempted to forget that Maria forgot?

In this context, more than twenty years later, Stewart's (1996) provocations for ethnography remain both timely and virtually unaddressed, especially in tourism studies (for notable exceptions see Little, 2010; 2014; Saldanha, 2007; Ness 2011; Crouch, 2010; Satsuka, 2015). "What," she asks, would happen "if [she] tried to arrest the progress of truth claims that reduce 'anecdotal evidence' to a secondary or deeply suspect status? What . . . [would happen] . . . if the density . . . [of things] . . . were allowed to disrupt intellectual concentration to a point where their material substance rubs against thought with a friction that generates cognitive sparks (Buck-Morss 1989 in Stewart 1996)? What if in the place of a transcendent system or code there was only the anecdote, the fragment—insufficient and unfinished" (Stewart 1996: 71–72)? For Stewart (1996; 2007), then, to unforget is not to repair a type of amnesia (for this is not amnesia at all), but rather, a politicized scanning and reattention "that displaces the rigid hierarchy of "subject" and "object"" (74); a type of "punctual reorienting of the event [that] . . . activate[s] . . . by a difference in register . . . new modes of encounter and creates new forms of life living . . . asking at every juncture what else life could be" (Manning 2016: 8).

Reenter Maria.

Again, what is so intriguing about Maria's inability to remember, whether as part of some strategy unbeknownst to me, or not, is that fact that it traffics purely in intensity. Here, the unspeakable parts of her speech act press on the "edges of life-living . . . with a force that . . . cannot yet be perceived, but [is] nonetheless felt" (Manning 2016: 53). That's affect, and as an intensifier, Maria's "not remembering" continues to carry the force of impact, instantiating further amplifications and reverberations with a scatter force in her life and the lives of those around her. In this sense, Maria's "forgetting" is not lost to the ether, but, rather, acts as a fragmentary force that intensifies the unfinished and durative *event-time of lingering*, a slow death yet to arrive, enacting a type of middling; a gesturing toward a "something happening" that is deeply registered in the body—felt but not yet perceived.

This is what Manning (2016) calls a "*Minor Gesture.*" "A continual variation on experience . . . [and that which] gets cast aside, overlooked, or forgotten in the interplay of major chords" (1), the minor gesture "invents new forms of existence . . . [those] temporary forms of life [that] travel across the everyday, making untimely existing political structures, activating new modes of perception, inventing languages that speak in the interstices of major tongues" (2).

Manning continues:

> As the activator, the carrier . . . [the minor gesture] . . . is the forward-force capable of carrying the affective tonality of nonconscious resonance and moving it toward the articulation, edging into consciousness, of new modes of existence. (7)

In this context, it matters very little whether or not Maria can actually remember if she said those things. It matters very little whether or not Maria was invoking some kind of strategy that she wouldn't tell me about. But, it certainly *matters,* in a literal sense. As a *minor gesture*, left to proliferate in "a mode of unforgetting" (Stewart 1996: 75), Maria's "not remembering" feels less like a gap in memory, less like amnesia, and more like affect—an intensity and potentiality that is not intentional nor unintentional; that is registered in the body but not yet perceived. It is more akin to what Little (Forthcoming) means when he says: "it's like looking at what's right in front of you but outside of your field of vision." To abbreviate this form would be to do an injustice to Maria.

Here, through lingering, through her endurance in the shocking and horrifying atrocities of colonialism, racism, and the neoliberal tourism-industrial complex in Costa Rica, the reverberations of which build the foundations of a touristic milieu forming in the aftermath of trauma and dispossession, living a slow death that has yet to arrive, Maria continues to enact an expressivity; a *worlding*. To forget that Maria forgot, to solely make "sense" of her outburst in terms of subjectification, that is, to measure the event-time of her endurance as a multiply interpolated lingering body, fragmentation and incompleteness notwithstanding, would be to miss the possibilities and potentialities of connectivity that her "outburst" activates in the liminal space of the encounter.

Much more than a site of resistance to a *good-life-tourism-machine* with recalibrative capacities to render her body as "troubled," "bullshit," or antithetical to some kind of imagined and invested *Pura-Vida-Paradise-Bliss*; much more than a site of confrontation that illuminates the stark realities and injustices enacted under the crushing and confining auspices of global tourism practices (although it is also intensely that); and, for my purposes, much more than an opportunity to "get the dirt" in the service of partial ethnographic truth claims, Maria's verbal tirade makes moves toward maneuverability, wedging open a something otherwise and a spark of hope. In this sense, Maria is more than an "object of oppression" or the "subject of resistance," but someone enacting the "haunting double vision of a subject who is simultaneously *in* a world and subject to it, and yet moving *through* it with the power to fabulate it and give it form" (Stewart 1996: 74).

This chapter, then, is an experimental attempt at ethnographically rendering this gestural fabulation. Not an attempt to finally "know" or "make sense" of Maria's outburst, but to mimic the durative capacities of lingering that Maria's forgetting activates, this is ethnography as a politics of extension of a moment of connectability, asking, at every turn, what else ethnography can be and what ethnography can do. Unfinished and fragmentary, it sits, left to resonate and proliferate, drawing from the refuse of what counts as a "good enough story of what's going on" (Stewart 2007: 4). *I can't remember*, Margarita said to me.

REFERENCES

Anderson, Moji. 2004. "Arguing over the 'Caribbean'" Tourism on Costa Rica's Caribbean Coast." *Caribbean Quarterly* 50, no. 3 (September): 25–46.

Berlant, Lauren. 2011. *Cruel Optimism*. Durham: Duke University Press.

Bruner, Edward. 2004. *Culture on Tour: Ethnographies of Travel*. Chicago: University of Chicago Press.

Clifford, James and George Marcus. 1986. *Writing Culture: The Poetics and Politics of Ethnography*. Berkeley: University of California Press.

Comaroff, John and Jean Comaroff. 1992. *Ethnography and the Historical Imagination*. Boulder: Westview.

Crouch, David. 2010. "Flirting with Space: Thinking Landscape Relationally." *Cultural Geographies* 17, no. 1: 5–18.

Deleuze, Gilles and Felix Guattari. 1988. *A Thousand Plateaus: Capitalism and Schizophrenia*. Minneapolis: University of Minnesota Press.

Fabian, Johannes. 1983. *Time and the Other: How Anthropology Makes Its Object*. New York: Columbia University Press.

Faier, Lieba and Lisa Rofel. 2014. "Ethnographies of Encounter." *Annual Review of Anthropology* 43 (October): 363–377.

Frohlick, Susan. 2013. *Sexuality, Women, Tourism Cross-Border Desires Through Contemporary Travel*. New York: Routledge.

Frohlick, Susan and Lynda Johnston. 2011. "Naturalizing Bodies and Places: Tourism Media Campaigns and Sexualities in Costa Rica and New Zealand." *Annals of Tourism Research* 38, no. 3: 1090–1109.

Geertz, Clifford. 1973. *The Interpretation of Cultures: Selected Essays by Clifford Geertz*. New York: Basic Books.

Gordon, Avery. 2008. *Ghostly Matters: Haunting and the Sociological Imagination*. Minneapolis: University of Minnesota Press.

Gregg, Melissa and Gregory J. Seigworth. 2010. "An Inventory of Shimmers." In *The Affect Theory Reader*, edited by Melissa Gregg and Gregory J. Seigworth, 1–28. Durham: Duke University Press.

Hemmings, Claire. 2005. "Invoking Affect: Cultural Theory and the Ontological Turn." *Cultural Studies* 19, no. 5: 548–567.

Little, Kenneth. 2012. "On Edge in an Impossible Paradise." In *Emotion in Motion: Tourism, Affect and Transformation*, edited by David Picard and Mike Robinson, 283–296. New York: Routledge.

Little, Kenneth. 2014. "Mr. Richie and the Tourists." *Emotion, Space, and Society* 12, no. 1 (August): 92–100.

Little, Kenneth. (Forthcoming). *On the Nervous Edge of an Impossible Paradise*.

Manning, Erin. 2016. *The Minor Gesture (Thought in the Act).* Durham: Duke University Press.

Marcus, George and Michael M. J. Fischer. 1986. *Anthropology as Cultural Critique: An Experimental Moment in the Human Sciences.* Chicago: University of Chicago Press.

Massumi, Brian. 1992. *A User's Guide to Capitalism and Schizophrenia: Deviations from Deleuze and Guattari.* Cochin: The MIT Press.

Massumi, Brian. 2002. *Parables for the Virtual: Movement, Affect, Sensation.* Durham: Duke University Press.

Mazzarella, Williams. 2009. "Affect: What Is It Good For?" In *Enchantments of Modernity: Empire, Nation, Globalization,* edited by Saurabh Dube, 291–309. London: Routledge.

Ness, Sally Ann. 2011. "Bouldering in Yosemite: Emergent Signs of Place and Landscape." *American Anthropologist* 113, no. 1: 71–87.

Ochoa, Todd Ramon. 2007. *Society of the Dead: Quita Manaquita and Palo Praise in Cuba.* Berkeley: University of California Press.

Palmer, Steven and Iván Molina. 2004. "Introduction." In *The Costa Rica Reader: History, Culture, Politics*, edited by Steven Palmer and Iván Molina, 1–8. Durham: Duke University Press.

Pratt, Mary-Louise. 1992. *Imperial Eyes: Travel Writing and Transculturation.* New York: Routledge.

Pratt, Mary-Louise. 1986. "Fieldwork in Common Places." In *Writing Culture: The Poetics and Politics of Ethnography*, edited by James Clifford and George E. Marcus. Berkeley: University of California Press.

Povinelli, Elizabeth. 2011. *Economies of Abandonment: Social Belonging and Endurance in Late Liberalism*. Durham: Duke University Press.

Rivers-Moore, Megan. 2007. "No Artificial Ingredients: Gender, Race, and Nation in Costa Rica's International Tourism Campaign." *Journal of Latin American Cultural Studies* 16, no. 3 (December): 341–357.

Romero, Andreas. 2015. "Image as Method: Conversations on Anthropology Through the Image." http://somatosphere.net/2015/08/image-as-method-conversations-on-anthropology-through-the-image.html, accessed August 14, 2015.

Saldanha, Arun. 2007. *Psychedelic White: Goa Trance and the Viscosity of Race.* Minneapolis: University of Minnesota Press.

Satsuka, Shiho. 2015. *Nature in Translation: Japanese Tourism Encounters the Canadian Rockies.* Durham: Duke University Press.

Shapiro, Nicholas. 2015. "Attuning to the Chemosphere: Domestic Formaldehyde, Bodily Reasoning, and the Chemical Sublime." *Cultural Anthropology* 30, no. 3 (August): 368–393.

Stewart, Kathleen. 2007. *Ordinary Affects.* Durham: Duke University Press.
Stewart, Kathleen. 2011. "Atmospheric Attunements." *Environment and Planning D: Society and Space* 29 (September): 445–453.
Stewart, Kathleen. 1996. *A Space on the Side of the Road: Cultural Poetics in an "Other" America.* Princeton, NJ: Princeton University Press.
Taussig, Michael. 2004. *My Cocaine Museum.* Chicago: University of Chicago Press.
Thrift, Nigel. 2008. *Non-Representational Theory: Space, Politics, Affect.* New York: Routledge.

NOTES

1. A denunciation. Or an official statement/accusation regarding the committing of a crime in Costa Rica.

2. A criminal investigation unit of the Costa Rican police force. What people often refer to as the secret police.

3. A small village approximately 20 kilometers from Puerto Viejo which houses municipal offices as well as larger police stations such as the OIJ.

4. Snapper refers to a 1,000 Colones note, approximately two dollars. It is called a snapper because of its red coloring, in specific reference to Red Snapper fish, which are common in the ocean neighboring the town.

5. This conceptualization of habit is very particular. In Massumi's (2002) work, the "self" always retains a hyphen (e.g., "Self-"), so as to indicate that "the self is not a substantive but a relation" (14).

6. Here, I use "abstracted" in the context of the theoretical argument I have been building. Maria's suffering, like Oskar's, is anything but abstract.

7. A 325ml bottle. Similar to a "Mickey."

8. Also referred to as Cacique. An inexpensive and locally available cane alcohol.

Chapter 7

"The Baskets Cannot Send the Children to School"

Women, Handicrafts, and Tourism in Botswana's Okavango Delta

Frances Julia Riemer

Women in Botswana's Okavango Delta weave some of the most beautiful and complex baskets in the world. For over two decades, their baskets were sold in tourist shops throughout the country, and images of the baskets were on postage stamps and telephone cards. The baskets also became highly collectible. One, woven by a member of the Weavers Group, is in the Smithsonian Museum's permanent collections. According to the Smithsonian's Curator of African Ethnology, the basket was "excellent and finely made," an example of the transformation of weaving from "baskets made by women for local consumption to new national and international art and craft."[1]

Two other baskets woven by fellow weavers in the group—one large and jar shaped, the second, bowl shaped and decorated with animal and geometric shapes—are housed in the permanent collection of the British Museum. All three baskets were donated by Anglo American PLC, a London-based international company that mines diamonds in Botswana, as part of the company's "commitment to supporting communities surrounding its operations around the world." According to company correspondence, Anglo American partnered with the Botswana Government, the Botswana National Art Gallery, and community craft organizations to launch the donations.

I first encountered baskets from the Okavango Delta in tourist shops in Gaborone, Botswana's capital, upon my arrival in country in 1995. Just about a month later, I met the weavers in a church compound in the Delta, the center

Figure 7.1. Basket in the Smithsonian Museum's permanent collection.
Source: Photo by Frances Julia Riemer.

of the country's basket weaving. Trained as an educational anthropologist, I sought out the weavers when I observed Thimbukushu[2] literacy classes facilitated by Fellowship, an organization of Christian churches, in the mid-1990s. I wanted to know more about the women's literacy, design, and weaving skills and their navigation of the local tourism and global marketplaces.

In my own visits and conversations with Botswana's weavers in the Okavango Delta over the past 20 years, I watched women, alone and in groups, weave baskets like those donated by Anglo American. I observed their attendance at and facilitation of workshops in weaving techniques, and watched them sell their baskets to Fellowship missionaries and later to representatives of Culturecrafts, "the largest retailer and exporter of handicrafts[3]" in Botswana. The weavers used fiber *mokola* palm fronds, colored with natural dyes from boiled roots and bark. They wound the palm into coils, and then wove the coils together with an awl, a long-pointed spike that pierces small holes in the coils. You can identify baskets from Botswana by the coil, tight weave, muted colors, and animal designs—tears of the giraffe, forehead of the zebra, knees of the tortoise.

During the best of times, the women's baskets supported the village's commercial enterprises, the work of the Fellowship, and Botswana's tourism handicraft industry. But during my last visit to the Okavango Delta in 2018, women were no longer weaving. A tourist handicraft industry that began as refugee relief in 1970, became part of Botswana's national branding, supported both missionary and regional tourism activities over three decades, and brought baskets to the Smithsonian and the British Museum, was, if not dead, then in a coma.

In this chapter, I tell the story of the rise and decline of Botswana's basket initiatives, and relatedly, contribute to a conversation about women, handicrafts, and tourism. I contend that the weavers' experiences can be best understood through postcolonial feminist theory. When viewed with that lens, the women's weaving project was untenable by design, despite the beauty and skill of the women's work. In this telling, in order to "save" women, refugee workers and expatriate missionaries monetized women's work, as described by Maria Mies as *housewifization* (1968,106), in a business model dependent on an external marketplace and unsustainable natural resources.

Let me say upfront that this is not the story I want to tell. Drafting this chapter has been both challenging and transformational. As a former development and refugee worker focusing on women's projects in east and southern Africa, Southeast Asia, and Latin America, I have long been committed to the idea of improving women's lives through education and increased income generation possibilities. Since I was introduced to weavers, I have been an advocate of their work. Along with wildlife safaris, the baskets served as identity for Botswana's tourism trade. My previous experience in adult education also made me believe that the challenges of crafts projects could be addressed with additional skills, management, and marketing training. Moving from contemplating project improvements to questioning the development-oriented model itself has been both complicated and at times, disorienting. Post-colonial feminist thought has provided a timely, nourishing, and much needed heuristic in reframing the presuppositions around the Weavers project specifically, and I suggest, women's crafts in the tourism market more broadly.

HANDICRAFTS

Across the globe, women and female-headed households experience higher rates of poverty than their male counterparts. At least since the 1970 publication of *Women's Role in Economic Development* (Boserup 2007), women's poverty and that of their families has been identified and named as a problem. But even before the United Nations made women's poverty an official problem, schemes—from micro financing to skills training to community-based projects—were widely acclaimed as solutions to women's economic challenges. Fostering women artisans has been a popular economic development strategy of both governmental and non-governmental organizations, with the World Crafts Council (WCC) founded in 1964 to help economic growth and to encourage craftsmen across the world. In blogs and websites (Indego Africa 2016; Sasibal 2016), we are told that the global artisan economy is a $34 billion per year industry, and the Global South accounts for 65 percent of handicraft exports around the world.

Arts and crafts have long been considered integral to tourism. Crafts, in the form of souvenirs and mementos, "spatially and temporarily expand the visitor experience" (Phillips, Indar, & Thorme 2017, 5). Souvenirs make the travel experience real to the tourist, and to friends and relatives who remained at home. I draw on Wilkins (2011, 5), who asserts that the purchase of crafts is tied to the visitor's need to preserve his or her experience as a "sacred, extra-ordinary time and space." In terms of economics, the OECD (2005) posited that linking tourism and crafts can both develop handicrafts and invigorate tourist destinations. The first International Conference on Tourism and Handicrafts, held in Tehran, Iran, in May 2006, focused on connecting tourism and handicrafts in poverty alleviation and on the role of local tourism in the preservation of traditional crafts (World Tourism Organization 2008). In fact, in many countries, handicrafts are seen as a main core of tourism development, and tourism facilities are often established near handicraft production centers.

Botswana's tourism is based in its eight national parks, game reserves[4] and wildlife management areas, which are home to Africa's Big Five: lions, leopards, buffalos, rhinos, and elephants, in addition to giraffe, brown hyena, warthog, wild dog, cheetah, red lechwe, kudu, and springbok. The country's tourism model is one of high quality / low volume, which translates to bucket-list packaged hunting and photographic safari experiences for upper middle class, mature travelers. Baskets and other handicrafts—mopane wood carvings, clay pottery, and "bushmen" art (e.g., ostrich eggshell beadwork, arrows, and hunting sets) are sold in safari camps, in gift shops in Gaborone, the country's capital, and in Francistown, Maun and Kasane, all gateway communities to national parks.

I'm a big fan of handicrafts, what curator and writer Glenn Adamson (2018, 28) calls *materiality* of an object. "When we say that someone crafts an object," writes Adamson, "we mean that they put their whole self into it, body and mind alike, drawing on whatever skills they have acquired over the course of their lives. This is a deeply meaningful human activity." Anthropologists have a long-standing interest in handicrafts, defined by UNESCO (2019) as artisanal artifacts made mainly by hand, generally using raw materials from sustainable sources. Classic anthropological theory focused on the role, meaning, and exchange of cultural commodities in social life (Malinowski 1922; Maus 1954; Douglas & Isherwood 1979). As part of study on material culture, anthropologists have been interested in the craft economy, including the acquisition of raw materials and the distribution and use of created products (Costin 2005). Exploring the relationship between aesthetics and the wider environment, anthropologists study craft as cultural design and identity (Cone 1995; Nash 1993; Wilkinson-Weber & DeNicola 2016) and been involved in the revitalization of craft skills across the globe (Grimes & Milgram 2000).

Development scholars, grounded in anthropology, social geography, and political science among a range of other disciplines, positioned handicrafts as income generation for women and their families (Tanga & Maliehe 2017). As part of a variety of approaches to alleviating poverty, scholars have studied women producers, the fairtrade craft industry (Hutchens 2010), and the efficacy of the handicrafts sector in the broader economy (Mustafa 2011). The growth of handicrafts has also caught the attention of anthropologists of tourism, who view the articulation of domestic and subsistence activities as embodiments of culture in motion (Appadurai 1988). Their work focused on the effects of mass tourism on local handiwork and cultural identity (Little 2008; Nason 1984; Swain 1993).

As I illustrate in the following sections, the story of Botswana's basket weavers is familiar when told through both anthropological and development lenses. In telling the history, I recognize and commend the intent of the project designers, trainers, and marketers, and cheer for the increased visibility of and respect for the women's work. But feminist anthropologists bring a very different orientation, focusing on crafts as representation of gender relations (Brumfiel & Robin 2008; Messick 1987) and as alienated home-based labor (Mies 1983). Prugl (1999, 78) argued that newly independent states linked the fostering of house-based crafts to the promotion of tourism, deliberately cultivating both industries as part of nation building, while confining women to secondary status as "supplemental income-earners."

The weavers' story is the trajectory from palm leaf to women's hands to two universal museums was woven from colonial uprisings, the mucking about of missionaries in Africa, the best intentions of refugee and development workers from the Global North, the not always good intentions of traders, local and otherwise, and the craft of poor women, mostly refugees, others from minority groups in southern Africa, who weave to buy school uniforms and pay school fees for their children. By following this history, we get a glimpse at how colonial ideology of the savior and saved, modern and traditional, men's and women's work, consumer and producer, guest and host—are gendered in the tourism craft market. Despite best intentions, enhanced skills, and the emergence of a "new national and international art and craft," weavers in Botswana told me that even in the best of times, "baskets cannot send the children to school." Now, in these worst of times, the women are no longer weaving.

COLD WAR PROXIES AND SAVING REFUGEES

The weavers came to Botswana as refugees from southeastern corner of Angola, uprooted by Angola's proxy war in 1969. Three thousand two hundred thirty Hambukushu men and women fled the fighting and claimed refugee status in Botswana from 1967 to 1969. The Okavango was already home to a first wave of Hambakushu residents, as well as those of ethnic Bayai peoples and the Batawanas, the officially recognized major tribe that controlled the area of northern Botswana spreading out from Maun. According to the Motswana historian Thomas Tlou (1976). Hambakushu and Bayei pasturalists and horticulturalists had been in the area since 1750. Chief Letsholatbe, the *Kgosi* or Paramount Chief of the Batawana "adopted" the new wave of refugees and allocated 269 kilometers of land for them on the western edge of the Okavango Delta.

Despite the new home, the first generation of refugees hadn't changed their sense of selves. They are Bantu speaking river people originally from the Zambezi River area of central Africa, who lived by fishing, dryland farming, and growing corn, sorghum, millet, ground nuts, citron and beans. When they talked to me about what it means to be Hambakushu, they continued to talk about fishing, subsistence farming, and cattle. Dryland farming is difficult however, in a place where rains are unpredictable, and to offset the precariousness of farming in their new home, they exploited the dry lands for grazing, firewood collection, and fruit harvesting, and the wetlands for fishing and the collection of palm for basket making, thatching materials for housing construction, and plants for dyes and medicine. But equally relevant, the refugees were destitute, and according to Botswana policy, ineligible for work permits. The United Nations High Commission for Refugees (UNHCR), the Red Cross, World Food Programme (WFP) and the World Council of Churches (WCC) all took notice. Attending to income generation became a primary focus.

The WCC Resettlement Officer thought Hambakushu handicrafts—baskets, carved wooden animals, knives, drums—might appeal to tourists. He sold a few items to Botswana Game Industries (BGI), and after several false starts, connected with Culturecrafts, at that time a parastatal developed by the government's Botswana Development Corporation (BDC) to market handicrafts. He kept 10 percent of each shipment's price for basket buying operations, and used basket profits, along with funds from the All Africa Conference of Churches and the American Embassy Self Help Fund and a loan from the Cooperative Development Trust, to build and stock a cooperative that sold fresh and canned foods, household items, and clothes. Since 1972, the coop was one of the first buildings visitors saw when entering the village. The basket industry and cooperative were symbiotic. The baskets helped build the coop, and village residents shopped in the coop with money earned from weaving baskets. Prices were affordable, and the cooperative hired locally and paid well. The cooperative's profits were invested in a gasoline station and a separate clothing and food store. For at least three decades, the ventures went hand in hand.

The basket venture also worked well for Culturecrafts, which became a major actor in this story. By 1983, Culturecrafts was purchasing and reselling nearly 90 percent of all crafts made in Botswana (Terry 2001). With the resettlement agent and Culturecrafts creating a market for baskets, Bayei weavers, river people known for their fishing and canoe building technologies, joined their Hambukushu refugee neighbors in the enterprise. The scene was ripe for outside consultants, and over the next five years, from 1972 to 1977, a slew of technical advisors came and left the area. They assessed market potential, surveyed the weavers, coined names for basket designs, created a buying system that involved measuring for size and quality, printed a poster, tried to convince the weavers to make mobiles, coasters, and trivets, identified craft leaders, formed production groups, cultivated trial plots of mokola palm trees, experimented with alternative and chemical dyes, and formed a cooperative among Bayei weavers in a neighboring village. By the 1980s, expatriate volunteers were everywhere organizing production and marketing networks. They came from the Peace Corps, from the International Volunteer Service (JVS), and the Netherlands Development Organisation (SNV). Others were independent consultants: experts in handicraft development and marketing.

Girls began weaving as early as the age of 12 by watching their mothers, playing with knots, fashioning the straw into coils, and wrapping coils in circles until a flat basket emerges. An award-winning Hambukushu weaver, told me:

> I learned when my mother was making baskets. We just collected the palms adults thought were the worst and just tried to make a basket with them. You just tried. I watched my mother. After I tried myself, showed some interest, then my mother would give me advice. She told me to make baskets and sell them, because if there's a drought it's the only way to make money. My mother didn't tell me how the patterns should be. When I knew, I decided to make new patterns, just the way I wanted to make baskets.

THE HEYDAY OF THE BASKET PROJECT

A Dutch couple (with their very young sons) arrived as missionaries with the Fellowship in the Delta in 1990. The couple stayed for nine years and quickly became a new driving force behind Hambukushu baskets. They were my hosts during my first several stays in the Delta, and they put considerable weight into promoting the region's crafts.

Jan and Elisabeth[5] were Dutch Reformed of the Calvinist church whose influence in southern African traces back to the Dutch East India Company of the sixteenth century. Their mission was to be self supporting: they would receive no funds from the group's central office in Gaborone. Jan was a minister, and he acted as idea man, advocate, facilitator, trainer, business manager, and pastor, growing and overseeing the church, pre-schools and Thimbakushu[6] literacy classes, wholesale and retail basket sales, brick making, welding, and carpentry businesses, guest houses, a pottery, a bakery, a palm tree nursery and plantation, and a small cultural museum. Elisabeth ran the basket buying side of things, in addition to home schooling her boys, who were, as I wrote in my fieldnotes, "all white blonde, ranging from 5 to 11, reddened from the sun and dirty from the sand."

Elisabeth told me that after a year of Thimbukushu language classes, she stumbled into what would be her role for the next nine years, invigorating the area's basket production and sales. This was virgin territory for Elisabeth. She explained, "We're Dutch. We didn't know about baskets. When we look at baskets we think about what to put in them." Over their nine years, she made baskets a business. She organized training to encourage "different dyes, different patterns," and nurtured a corps of weaving teachers. She held a basket-buying day once a week at the Fellowship compound and once a month in Bayei villages to the north. Elisabeth bought baskets; Jan moved the baskets on to Culturecrafts in the capital, to safari companies and resorts, and out into the larger world of basket collectors. The profit from buying and selling baskets financed the other start-ups, and even paid 25 percent of the couple's salary. Jan and Elisabeth were everywhere, known to everyone in the region.

Women all over the Delta were weaving baskets, in their compounds, under trees, and in other slices of shade. "Our husbands like the basket making," one woman told me. "It brings in extra money. We farm, but we can start weaving by 11 when the sun gets hot." Weaving had become part of what anthropologist Louis van Tonder (1966) called a "risk aversion" economy, a combination of seasonal income generating activities of farming with beer brewing, basket weaving, plant collection, and work outside the area. Weaving complemented farming, cattle rearing, beer brewing, the sale of fat cakes, and work in the government's drought relief program. Weaving was seasonal, following the farming cycle. According to Elisabeth, "In July, August, [after harvest] there are few baskets, even though the demand for baskets is high. In January, February, our storerooms are overflowing." She continued,

> It also has to do with drought relief. If the government puts a lot of money into drought relief,[7] people will stop weaving. We try to encourage our best weavers to weave all year round. We think of them as artists, they are making art.

Over 1,200 women in the area sold baskets to the Fellowship. In 1995, the first-year marketing records were maintained, the church group bought 25,000 baskets from women who walked to its compound each week on basket day in the hope of selling something. The women waited from early morning for Elisabeth to set up her buying table for the basket buying at 2:00 pm. Elisabeth had trained two Hambakushu women, and together they measured size and quality, and gave each woman cash on the spot for her baskets. They evaluated the baskets, one by one, for the evenness of the coils, the color of the dyes, the intricacy of the patterns, and overall shape. There were five categories of baskets, going from low to high:

1. Ts, sold only in in the village,
2. Ps or premiums,
3. IMs or intermediates,
4. SPs or super premiums,
5. and CCs, a new highest caliber.

The price paid to the women increased between 43 percent and 72 percent from one category to the next, and doubled between an SP and CC.

On basket days, most women found their baskets rated as Ps. Out of 100 baskets that Elisabeth purchased on a single basket day, more than 90 percent were Ps. She rated only five as IMs, two as SPs and one CC. The weavers didn't like the outcome, but they agreed that in theory at least, Elisabeth's pricing structure was fair. It was the scientific aspect of the measuring device that gave the women confidence in the churchwoman's objectivity. "It's fair, Elisabeth uses a tape measure," the women asserted. But in less formal conversations, they complained about their lack of control over the market. As one of the region's best basket weavers explained at the time:

> No, we don't understand [decisions about prices]. So many people take different baskets there, which have different methods of weaving, and they get different ratings. I took one last Saturday, I thought it was an IM, Elisabeth thought it

Figure 7.2. Basket buying, Etsha 6, Botswana.
Source: Photo by Frances Julia Riemer.

was SP. When we make baskets we think are good, Elisabeth says they're not good. Sometimes we think they're not good, and Elisabeth likes them. Maybe the patterns we think are good, but when she looks at them, she says they're not in a good position.

On occasion, the women did protest the rating their basket was given, but Elisabeth never conceded. Power wasn't shared between buyer and weaver. The church group was then the sole basket buyer in town. Without another accessible marketplace or transportation, continuing to work with Elisabeth was the most rational and financially viable option.

What made the women's situation even more disconcerting is while they remained relatively powerless in the commerce side of the project, their products became more complex, beautiful, and expensive. Traditional Hambukushu baskets had been plain white, without design or colored dyes. Over the 25 years of training and prodding by consultants and NGO workers, the women weavers amassed over 30 designs taken from the world around them and borrowed from their Bayai neighbors. The quality of the weave improved due to Elisabeth's deliberate attempts to help the women advance their craft. As their sole point of sale, the women took her critique seriously. On basket day, Elisabeth gave feedback to each woman as she had her turn at the head of the selling queue. Across a single half-hour, Elisabeth told one woman she should pay more attention to the consistency of her patterning, talked to another about the inconsistency of her weave, pointed out to a third that the dye she used didn't go through to both sides of the basket, and showed a fourth that her big closed basket was too bumpy. In addition to this informal feedback, Elisabeth coordinated regular training sessions for interested weavers. "We started some training," she explained, "to encourage them to use different dyes, different patterns." Elisabeth continued,

> We started with those weavers, then we added some, and others quit. The teachers are good weavers. They know they are good weavers, they get high prices for their baskets. Some are doing well with traditional patterns, some like to go on to different things.

Elisabeth was explicit about her efforts to push the women, asserting, "So then I look for new things. Maybe I'll find a book on Native American baskets." She was creating baskets for a world market, driving the transformation of craft from utility to art.

Everyone agreed that the best weavers were the younger women. Without children of their own, they had fewer obligations at home. They could afford to work slowly, slicing the palms thinly, dying them evenly, and piercing them carefully with a well-sharpened awl. A master weaver explained that weaving a high-quality basket takes time.

> I try to remind others about making baskets. They say it takes too much time to make a basket like that. Yes, like me, I'm very slow. If I begin with someone making the same basket, that person will make three baskets while I do one. One

> I made, it too me a whole month. It is a CC. They gave me 150 pula ($37.50 US). I have to untie most of the time and redo it.

She calculated that she took three times as long as other weavers to finish a basket. But for most women, living at subsistence level meant not affording that extra time, even if it meant more money in the future. There wasn't the luxury of investing in the long term; the sense of immediacy was a function of poverty.

Yet the consensus among both collectors and the international development folks living in southern Africa was that these best baskets had progressed in quality. Starting as tourist souvenir, they were becoming art. A widely distributed Botswana basket poster, developed in the consultancy boom by Michael Yoffee, describes this transformation in terms far more poetic than my own. "The humble material of palm leaf is ennobled by the creative genius of Botswana's artisans, transforming the palm into an art form, ranking it with the great cultures of Africa."[8]

Starting in 1990, Botswana's National Museum in Gaborone, the capital, held a yearly basket exhibit, where baskets were exhibited, judged, sold, and auctioned for high prices. The exhibit was a ticketed affair. The hall was two storied, and both up and downstairs were packed with mostly Anglo expats. For the expatriate basket aficionados, the night was heaven. Beautiful baskets, nearly all from weavers in the Okavango Delta, were hung on all the walls, and closed baskets were standing on the floor. One wall contained the tendered baskets, all SPs, the highest category of quality. These were the most intricate of baskets by the best weavers, clustered by weaver. Each category of baskets, flat, open, and closed was judged, and the winners had been awarded ribbons. Enlarged pictures of several weavers were hung on the walls; I recognized nearly all of the women.

In 1996, the first time I attended the exhibit at the Museum, one weaver received 2500 pula, or $625 (at P4 per $1.00) for a large closed basket. The others, about 80 percent of the exhibited baskets, were IMs, the second category of quality, and were being sold on the spot. I bought a small closed basket in different shades of brown for 104 pula (then US $26), which was about twice the Fellowship's selling price and four times what the women made selling to the church group. I saw expatriate friends, as though hypnotized, spend hundreds of dollars on baskets. For two hours in the museum's exhibition hall, the desire to purchase became all-consuming.

A few of the weavers traveled to the exhibits on the government's dime. They spent the evening weaving, seated on the floor, their legs stretched out front, with a small coffee tin of water to dip fronds in between them. They were living cultural artifacts, the real thing, and the expat collectors were interested in everything about them. One of the weavers said people asked her,

> What makes basket rolls, palm and grass? Where do materials come from? How far for dyes? Are they in the Delta or on land? How many times a month do you go? Mostly the whites asked. Only a few Batswana. Maybe the whites want to know how baskets are made.

None of the women had ever traveled the 12 hours by road to Gaborone before the exhibits. They had never seen two-story buildings, elevators, traffic, or traffic lights. But beyond the wonders of urban life, for the women, the trip meant real money. A weaver offered, "It was my first time in Gaborone. I got 120 pula [US$30) for a small basket, I would have gotten maybe 50 something (US $12.50) here."

Things were truly going well for the weavers. Basket buying was predictable, and weavers were making good, steady money. The relationships among the missionaries at the Fellowship's headquarters in the Delta, the village cooperative, and Culturecrafts were productive and dependable. If anything, the concern was that the weavers couldn't keep up with demand. The seasonal nature of the women's weaving translated to the need for a sizeable inventory to meet orders from in and outside the country. The Fellowship's supply closet, where baskets were stored, was always locked and closely monitored. But while demand actually gave the weavers some power, they didn't take it. The project was dependent on the weavers, but they, in turn, felt dependent on the project. Maybe it was the tight control maintained; everyone knew that the church group was the sole buyer. I certainly knew. When I was in the Delta, I rarely bought baskets directly from the women. It would have been like betraying my hosts, invading the territory they had carved out as their own.

The Fellowship's work was vital and viable. The missionaries built guest houses, a new church and attached dining hall, they explained for "guests who'd like a simple meal when they stay in the chalets." A large arch had been built as an entryway to the compound, because, as they continued, "we thought we needed some nice entrance." While over time the ten guest houses became tarnished—chairs had broken, the baskets that had been hung on the walls as decoration had disappeared—the ablution block offered warm showers when the guard built a fire by the hot water pipe.

The missionary couple prepared to leave in 1998. Because expatriate departures are often the death of development efforts, they offered to sell the basket business to Culturecrafts, without telling their sponsor, the Fellowship. A legal advisor had instructed them against the sale, and the arrangement didn't work out. Instead, they identified local men and women to spearhead each of their projects, and left the oversight to the Fellowship office in Gaborone.

It didn't take long for things to fall apart. Administrators at Fellowship's head office in Gaborone expected its field office in the Delta to remain self-supporting. They did not contribute financially to fieldwork, and without financial contributions, work there ground to a halt. The Fellowship compound in the Delta was left without a manager and a minister for nearly ten years, and the two or three employees who had been trained by the missionaries struggled to make ends meet. The guesthouses remained open for a few years, and private toilets were added to each chalet to accommodate tourists and Batswana officials traveling from the capital to the bush. But without money, water and electricity couldn't be paid, repairs couldn't be made, roofs couldn't be re-thatched. The houses leaked when it rained, and furniture and bedding became mildewed and torn. In 2008, headquarters began a dispute over taxes with the government, and the guest houses were closed completely. In just a few years, the compound grounds became unkept, littered with empty Coke cans and beer bottles, and the small businesses were at a standstill. When the palms in the trial plantations were chopped and thrown over the fence by a council employee, no one had protested. No one seemed to care. What had been a museum and cultural center were in ruins. The roof had collapsed, windows were all broken, shattered glass covered the floor, the baskets had all disappeared, and from the empty beer bottles on the floor, it looked to have hosted occasional local parties. All that was left was a wooden canoe, a *mokoro,* in one corner.

Basket buying had the longest life. Buying baskets brought in money to support basic operations, but eventually that money dried up as well. Women continued to weave, but the Fellowship was no longer able to act as a reliable conduit to the outside world outside. Basket buying went from once a week to once a month to once every few months, and then only if an unexpected order for baskets arrived.

This post-missionary period was also a time of change in the country's economic health. Landlocked by white minority regimes of South Africa, Namibia (then South West Africa), and Zimbabwe (then Rhodesia) during its neighbors' wars of independence, Botswana was viewed by the global community as both a safe hold and a safe investment. When apartheid in South Africa finally ended in 1994, the region opened and many of Botswana's expatriate development workers were re-assigned to other countries. Then in 1996, the World Bank classified Botswana as an upper middle-income economy,[9] good news for the country's financial markets, but bad news for industries that depend on external consumers. It was expatriates in Botswana who bought the women's baskets, people working for the government or non-governmental organizations. With the change in World Bank status, they left Botswana, following the donor dollars to poorer parts of the globe. Fewer and fewer people remained in country who were interested or able to purchase baskets that cost more than a hundred dollars a piece. And connecting to safari tourism and tapping into external markets was a feat well beyond the weavers.

THE BEST OF THE BEST

At that point, Culturecrafts's Managing Director developed and pursued a different business model for selling baskets. He gathered the best weavers to weave everyday and called them the *Weavers*. In 1998, the group was composed of 11 weavers and over the next 14 years scaled up to 24. Culturecrafts paid them a wage, 400 pula, or just about US$50 a month, to weave, Monday through Friday, 8:00 am to 4:00 pm. The women produced highly detailed baskets without a missed hook or errant weave. Culturecrafts bought the baskets, and sold them in tourist sites in Botswana or overseas for prices from $100 and up. Baskets of extraordinarily intricate weave were hung on walls and propped up on stands in the back room of Culturecrafts's main shop in Gaborone. The organization also developed a webpage dedicated to the weavers, who were all identified and introduced to the larger world, and produced a *Botswana Baskets* book of photographs and text with the help of one of the earliest crafts consultants to the area. In addition to weaving their own baskets, when Culturecrafts needed inventory, the women bought baskets from other area weavers who had moved to Maun, the largest town in the Delta,[10] following directives concerning what kind of baskets—open, closed, large, small—to weave and purchase.

I had followed this development closely: I was very interested in what it meant to the women to be the best basket weavers in the country, finally paid to weave full time. On one of my return trips to Botswana, I visited with the women during several of their 9:00 am to 4:00 pm days of weavings. They sat on the ground in the yard under a large acacia tree and on a covered porch in the back of Culturecrafts's compound. They wove and chatted about the weaving, about life in the Delta, about friends and family. Moving from weaving baskets to make money to weaving art from mokola palm meant something to me, and I wondered what it meant to the women themselves. I was surprised when the women complained that the work was difficult, that weaving hurt their eyes, their backs, and was basically dead end. "Has no future" are the words they used. They wished aloud for funds to start other businesses—tuck shops for instance, which surprised me given the high quality of weaving in which they engage. But their concerns were rational. Each basket takes to several weeks to make, and weaving is backbreaking work. Four hundred pula or US$50 a month is not much in Botswana, where, according to the World Bank, the GNI per capital income in 2011 was US$7,470.

The weavers saw themselves as employees. They liked the director, telling me, "he helps us with contributions in deaths of your daughter, your mother. He buys us tea, everyday we have tea. We are also paid leave over Christmas." But they balked when he floated the idea of their becoming shareholders in the Weavers. Buying stock made no sense to the women, even with the option of buyback after two years. They couldn't afford the longer-term investment. Six-hundred pula a month in hand and a higher price on each basket was a more understandable and manageable financial arrangement.

The main complaint was for their children. "We have children at home, they need uniforms," one woman explained. "When they graduate from form 3, there are no jobs, they stay at home. We'd like our kids to do something better than we are." Another added, "we want them to be nurses, teachers, be on the council. Have office jobs." Weaving, they explained, was a last choice. "Those who haven't done well in school, they'll have to weave," a weaver added. "We'd like our children to do better than we are."

A few entrepreneurial Batswana also tried their hand at the basket market. The accepted line was a desire to help the "poor weavers," but the reality was that money might be made on the baskets. Culturecrafts's profit per basket was about nine times what they paid the weavers, part of which included marketing, inventory management, and public relations responsibilities. A local woman adopted a similar business plan, hiring 20 other best weavers to weave full time for a monthly wage. She wrote on her website that her company's objective was "to empower Batswana in rural areas to improve the quality of live [*sic*] through employment creation, promotion of locally made handicrafts and utilization of locally available raw materials." I'd read about her in the press; she had become an active member of the Women's Alliance, and her name appeared in press blurbs when the Alliance held a luncheon. She told me her focus was on the export market, but that she didn't get very far because there was no export market for the baskets. She tried to establish a market in China through the Botswana Export Development & Investment Authority (BEDIA),[11] but couldn't get a foothold. The weavers, however, had less than positive things to say. They explained:

> She hired us. We made big baskets for her. She'd take them and not pay us. If she paid us, she'd give us 230 pula or something like that. She still owes many people.

The word was that the woman had also taken $2000 worth of baskets from the Fellowship's storeroom to sell, but had not returned to pay the organization.

I was hopeful when I learned about Anglo American's 2010 "initiative to help raise the international profile of Botswana basket making and craft in general."[12] Anglo American, the company that had a monopoly on Botswana's diamond mining, had not only taken notice of the weavers, but had donated baskets to the Smithsonian and the British Museum permanent collections. Unfortunately for the weavers, Anglo American apparently dropped the ball, and very quickly at that. According to an Anglo American spokesperson, no one at the company knew of any support provided to the basket weavers since the 2010 exhibit.

My last visit to Botswana found the Fellowship compound in the Delta shuttered. As I wrote in my notes,

> After three and a half hours on the road from Maun, I made it to the Fellowship compound, and am struck by its deterioration. Everything is locked, and weeds and cows have taken over. I am sad, remembering how active the compound was when I first arrived in 1998. Now everything is closed—the bungalows, the church, the bakers, the shop. As I took a panorama photo of the compound overwhelmed by weeds, I thought, this is what unsustainable looks like. A woman walking by greeted me. She called the former buyer, who arrived, gave me a big hug, unlocked the famed inventory closet, and showed me the remaining 22 baskets, covered in thick dust. I bought all 22 baskets; she took great care in calculating the prices by referencing the Fellowship's price list.
>
> I heard from two sources that the Fellowship resettlement officer died last year. I visited the village coop, and was equally struck by how empty and dark it looked. My heart feels a bit broken.

Maybe this is a story about my own naiveté. The Weavers project, born of refugee workers and cultivated by missionaries and a crafts parastal, had a happy and healthy life. I hadn't realized it would also have an end date.

MAKING SENSE

In Botswana, safari tourism continues to thrive. But the Fellowship no longer has funds to purchase baskets, and Culturecrafts, the big player for more than a decade, has moved in a different direction. The Weavers Group remains a link on the company's website, but Culturecrafts has been recast as a venue for regional music. A restaurant and concert venue, which regularly hosts African jazz and hip hop concerts, sits next to Culturecrafts's retail shops. It became one of the hottest places in town to listen and dance to local and international live music.

It is clear that the market for baskets made in the Okavango Delta had never been internal. Ever since baskets were commoditized in the early 1970s, the market had been external. And that was never more true than at the National Museum's Basket Exhibit. Not many Batswana attended the exhibit, since not many were interested in local crafts. A Motswana friend explained that few of her friends embrace both the "traditional" and big city culture. In Botswana, one's gaze is toward the city. New things show you are modern; baskets are viewed as local, traditional. They are the village, not the world outside the village. Historian Peter Stearns (2001, 119), in his *Consumerism in World History,* noted that in Africa, "consumerism, in other words, was a central issue in a war between change and tradition, between the individual and intense communal and family ties." This distancing from "the old world" is not unusual. Sociologists, anthropologists, and historians have written about experiences with modernity and consumerism. Steams (2001, 46) wrote that in the mid-eighteenth century in the US, "urbanites, in particular, were eager to distinguish themselves from country bumpkins. They did that by what they bought." Helena Norberg-Hodge (1992, 99) described the elevation of the modern and stigmatizing of the traditional in Ladekh, in northern India. Norberg-Hodge asserted that increased access to the western world, through tourism, what she referred to as "western-style" education, and media images, resulted in people complaining that the village was "primitive, silly, and inefficient." Over the course of her work in the region, she saw Ladekh residents rejecting their own culture for what is "modern." "Even the traditional foods are no longer a source of pride," wrote Norberg-Hodge. "Now when I'm a guest in a village, people apologise if they serve ngamphe instead of instant noodles."

In Botswana, the loss of marketplace connections for the women's high-end products, in combination with a domestic market focused on manufactured imports, has translated to both to economic and cultural loss. Post developmental theorists Susan George (1997), Arturo Escobar (1994), and James Ferguson (1990) help make sense of the project's slow demise. They argue that the basis of development initiatives in the Global South, that is, prioritizing financial growth based on western economic models, is harmful to both the land and its human stewards. The gaze toward the north, wrote Arturo Escobar (1994), dismisses local cultures and sensibilities. In the case of the women's baskets, the use of local materials and homegrown skills did not mean the gaze was anywhere but toward the North. It's not a stretch to argue that once baskets were transformed from a utilitarian container to collectable commodity, the project became unsustainable, dependent on an external marketplace.

Post-colonial feminists take this critique one step further. To promote the political project of decolonization, they forefront gender analytics in the unpacking of complex power relations between the colonizer and the colonized. They argue that in their eagerness to save "poor, traditional women" (Spivak 1988), white feminists overlooked racial, cultural and historical specificities of these women. Employing that frame here, the commoditization of the women's weaving originated in a white liberal feminist model of economic empowerment, which in practice reflects colonial relationships, roles, and representations, reinforces long-standing social hierarchies, and reproduces the very poverty, dependency, and disempowerment the project intended to solve.

The commodification of the Weavers baskets sustained the gaze from South to North, from the formerly colonized to a neo-colonial world. As I was taking leave of the Weavers, the women told me, "we would like you to help us. If there's money, some would start a tuck shop. Some would do something else, because this job has absolutely no future. We would like you to help us." And you know what? I scrambled to figure out ways to help. "Who can I connect," I asked myself. "Maybe I can hook up with safari companies or tourist hotels? Or start my own fair trade basket shop? How can I step in to save these women?" Yet somehow I suspected that none of that will change the way colonialism is encoded in a basket.

Normative research on crafts and tourism consumption has produced typologies of travelers shopping behaviors in order to improve the planning, marketing, and management of souvenir sales (Anderson & Littrell 1996; Masset & Decrop 2020; Silva et al. 2020). But when the focus moves from consumer to producer, the story shifts to one that more closely represents Mies's description of the integration of "Third World women's work into the global market" in the informal sector. Mies (1986, 115) wrote,

> It is characteristic of this sector that some of the products produced by the women were items traditionally produced for the consumption of the community, that is, they had no exchange value but only a use value. With the integration of such goods into an external market system, these items become commodities and the producers become commodity producers, even if they maintain the same form of production, for example, household production. In recent years, a deliberate effort has been made—known as the strategy to increase income-generating activities among poor Third World women—to link this area of women's work to the world market.

In Botswana, church groups and non-profits commodified women's weaving as high-end collectibles. Their products subsidized the operations first of the village cooperative, and then the Fellowship and Culturecrafts. By buying and reselling the women's baskets at an immediate 250 percent to 350 percent mark-up, the Fellowship, in its heyday, was able to support all its other operations in the Delta. In essence, the baskets afforded the Fellowship the opportunity to have a heyday. And by marketing the women's baskets at a 100 percent mark up from the price it paid to the Fellowship, Culturecrafts became an independent marketing entity. For a short time, baskets made for some good public relations for Anglo American. Anglo American's London exhibit got a small blip on the company's do-good report. But for a mining company with a controversial history in Botswana, including allegations of colluding with the Botswana Government in forced removals of indigenous peoples from the Central Kalahari Game Reserve in order to mine the land, the donation, albeit important in terms of the weavers and their handicrafts, was spin doctoring, not sustainable support and access to the market.

Feminist decolonial epistemology demands a shift in focus to the ways of knowing of the colonized. It emphasizes the multidimensional vectors of power that structure both lived identities and social reality, illuminates ties between epistemic location and knowledge production, and offers analytic strategies that linked the material, the discursive, and the structural. As the Santa Cruz Feminist of Color Collective (2014, 33) has noted, anticolonial feminist theory moves intersectionality beyond critiques of state-based legal practices to illuminate "glocalizing" dynamics—interpenetrations of global and local that construct gender, race, class, and sexuality, not as separable categories, but as mutually constituting systems of power that exist in and through contradictory and conflicting relations."

In Botswana, race, class, nationality, ethnicity, and clan membership complicate gender. While educated urban Batswana women have taken the lead in successfully contesting discriminatory marital property and citizenship laws (Allen 2015; Dow 1995), the country has also had a troubled history with refugees, dating to the government's policies and practices toward South African refugees during their country's liberation struggles (Allen 2015). The practices of the few entrepreneurial Batswana who have reached out to help the "poor weavers" suggest that actions along gender lines do not necessarily ensure increased equity and social justice.

The questions then—which criteria and political projects lead to decolonization? What counts as decolonization? And which practices succeed in challenging colonialism and coloniality?—still need to be addressed. What might crafts production and women's tourism initiatives look like though an anticolonial lens? The question of sustainability looms large. Mies (1986) also asserted that "feminist concept of labour maintains that work retains its sense of purpose, its character of being useful and necessary for the people who do it and those around them. This also means that the products of this labour are useful and necessary, and not just some luxuries or superfluous trash as are most of the handicrafts made today by women in 'income-generating activities' in Third World Countries." While Mies writes with a lens that focuses on capitalist patriarchy, her analysis is relevant to a feminist postcolonial analysis of Botswana's basket weavers.

Structurally, a business model of women's tourism work based on dependency—on an external market, on white mediators—has been a tenuous enterprise, for the women anyway. Assistance to rural and ethnic minority women from Botswana's central government has also been limited and unreliable, and government support of local tourism initiatives and cultural heritage has been uneven. Without re-centering from consumer to weaver, without a business model that forefronts sustainability, and without government positioning of handicrafts as integral to the country's tourism planning and cultural heritage, the project was bound to struggle.

Yet the development-based business model for women's handicrafts in the tourism market is far from unique. Stories of tourist handicraft coops supported by expatriate volunteers (Kaufman 2004), nonprofits (Horjan 2011), and church groups (Matsenjwa & Musiiwa 2016) appear in the popular press and in tourism literature. A postcolonial feminist consciousness reframes this model by highlighting central questions raised by Spivak (1988): Who speaks and for whom? Who is represented, by whom, and to whom? As designed, conceptualized, and operationalized, the projects cast white women from the global north as saviors, mediators, and consumers and the weavers as traditional "Third World" women, as historical subjects, and as poor women of Africa (Mohanty 1988). The models reflect, create, reproduce, and are dependent upon neo-colonial relationships and hierarchies.

By centering the women's experiences, I trouble the neo-liberal, neo-colonial development model on which the work of the weavers has been based. So what does this story tell us about tourism and handicrafts? Proffered policy recommendations involve extending the reliance on external markets by linking handicrafts to tourism activities, providing marketing channels for selling handicraft products, and marketing products internationally (Mkenda & Aikaeli 2019; Mustafa 2011; Wisudawati & Maheswari 2018). But the Weavers story suggests that for craftspeople, reliance on an external market can be both tenuous and challenging. Harkening back to Maria Mies, it's important to note that in terms of the supply chain, the weavers were only involved with production, that is, in obtaining and turning raw materials into handicrafts. They relied on, and it would not be a stretch to argue that in the process, they were used by, neo-colonial networks—in the form of foreign intermediaries—for both packing/storage and marketing and sales. Equally important, that dependence was unsustainable. While Botswana's tourism sector—the high-end safaris—continue to thrive, without dependable intermediaries, the weavers' connections to tourism, as well as to the external market, have frayed, and Botswana's basket handicraft industry is on the verge of disappearing.

REFERENCES

Adamson, Glenn (2018). *Fewer better things: The hidden wisdom of Objects.* London: Bloomsbury.

Aitchison, C., Macleod, N. E., and Shaw, S. (2000). *Leisure and tourism landscapes: Social and cultural geographies.* London: Routledge.

Anderson, L.F., & Littrell, M.A. (1996). Group profiles of women as tourists and purchases of souvenirs. *Family & Consumer Sciences.*

Appadurai, A. (1988). *The social life of things: Commodities in cultural perspective.* Cambridge, UK: Cambridge University Press.

Ateljevic, I, Pritchard, A., & Morgan, N. (2007). *The critical turn in tourism studies: Innovative research methodologies.* London: Routledge.

Bianchi, R.V. (2009). The "critical turn" in tourism studies: A radical critique. *Tourism Geographies* 11(4), 484–504.

Bircan, I, Ulker, H.I., Gunes, G., & Karakoc, G. (2010). Tourism destination sustainability and non-governmental organizations (NGOs): A case study of Beypazari, Turkey. *Journal of Educational Travel* 1(1), 17–32.

Boahen, A.A. (1985). *General history of Africa, VII: Africa under colonial domination, 1880–1935.* UNESCO, Heinemann Educational. London: University of California Press.

Boserup, E. (2007). *Woman's role in economic development.* London: Routledge.

Brumfiel, E.M., & Robin, C. (2008). Gender, households, and society: An introduction. *Archeological Papers of the American Anthropological Association* 18(1), 1–16.

Cone, C.A. (1995). Crafting selves: The lives of two Mayan women. *Annals of Tourism Research* 22(2), 314–327.

Costin, C.L. (2005). Craft production. In H. Maschner (ed), *Handbook of methods in archaeology.* Lanham, MD: AltaMira Press.

Douglas, M., & Isherwood, B. (1979). *The world of goods.* New York: Basic Books.

Dow, Unity. (1995). The citizenship case: The Attorney General of the Republic of Botswana vs Unity Dow: Court Documents, Judgments, Cases and Materials. Lentswe la Lesedi.

Escobar, A. (1994). *Encountering development: The making and unmaking of the Third World.* Princeton: Princeton University Press.

Ferguson, J. (1990). *The anti-politics machine.* Minneapolis: University of Minnesota Press.

George, S. (1997). *How the* poor develop the rich. In M. Rahnema and V. Bawtree (eds.), *The post-development reader* (pp. 207–213). London: ZED.

Grimes, K.W. & Milgram, B.L. (2000). *Artisans and cooperatives: Developing alternative trade for the global economy.* Phoenix: University of Arizona Press.

Hannam, K., & Knox, D. (2010). *Understanding Tourism: A critical introduction.* London: Sage.

Horjan, Goranka. (2011). Traditional crafts as a new attraction for cultural tourism. *International Journal of Intangible Heritage* 6, 46–56.

Hutchens, A (2010). Empowering women through Fair Trade: Lessons from Asia. *Third World Quarterly* 31(3), 449–467.

Indego Africa (2016). How the artisan sector can change the world. *One Website*. https://www.one.org/us/blog/how-the-artisan-sector-can-change-the-world/.

Jacques, G. (1994). Drought in Botswana: Intervention as fact and paradigm. *Pula: Botswana Journal of African Studies* 9(1), 34–60.

Kaufman, S. (2004). Handmade in the Turks & Caicos: The Evolution of the Middle Caicos Co-Op. *Times of the Islands*. https://www.timespub.tc/2004/01/handmade-in-the-turks-caicos-the-evolution-of-the-middle-caicos-co-op/.

Little, W.E. (2008). Outside of social movements: Dilemmas of indigenous handicraft vendors in Guatemala. *American Ethnologist* 31(1), 43–59.

Lorde, A. (2018). *The master's tools will never dismantle the master's house*. New York: Penguin Classics.

Malinowski, B. (1922). *Argonauts of the Western Pacific: An account of Native enterprise and adventure in the Archipelagoes of Melanesian New Guinea*. London: Routledge & Sons.

Masset, J., Decrop, A. (2020). Meanings of tourist souvenirs: From the holiday experience to everyday life, *Journal of Travel Research*, 10.1177/0047287520915284, (004728752091528).

Matsenjwa, G. N., & Musiiwa, E. (2016). The globalization of the rural Swazi weaver: Lavumisa women in the commercial handicraft industry, 1981–2013. *Journal of International Women's Studies* 17(1), 102–116.

Maus, M. (1954). *The gift: Forms and functions of exchange in archaic societies*. London: Cohen & West.

Messick, B. (1987). Subordinate discourse: Women, weaving, and gender relations. *American Ethnologist* 14(2), 210–225.

Mies, M. (1983). *The Lacemakers of Naraspur*. London: Zed Books.

Mies, M. (1986). *Patriarchy and Accumulation on a World Scale: Women in the International Division of Labour.* London: Zed Books.

Mohanty, C. (1988). Under Western eyes: Feminist scholarship and colonial discourses. *Feminist Review* 30, 61–88.

Mustafa, M. (2011). Potential of sustaining handicrafts as a tourism product in Jordan. *International Journal of Business and Social Science* 2(2), 145–152.

Nash, June C. (1993). *Crafts in the world market: The impact of global exchange on Middle American artisans*. Albany, NY: SUNY.

Nason, J.D. (1984). Tourism, handicrafts, and ethnic identity in Micronesia. *Annals of Tourism Research* 11(3), 421–449.

Nkomazanna, Fidelis, & Setume, Senzokuhle Doreen. (2016). Missionary colonial mentality and the expansion of Christianity in Bechuanaland Protectorate, 1800 to 1900. *Journal for the Study of Religion* 29(2), 29–55.

Norberg-Hodge, Helena (1992). *Ancient Futures: Learning from Ladakh.* Oxford: Oxford University Press.

OECD (Organization for Economic Cooperation and Development). (2005). "Culture and Local Development." OECD Publishing.

Parsons, N. (n.d.). Botswana. Encyclopedia Britannica. https://www.britannica.com/place/Botswana.

Phillips, W., Indar, D., & Thorme, E. (2017). An analysis of the arts and craft sector and its potential for sustainable tourism development in the Caribbean. *The Caribbean.* 63. United Nations ECLAC.

Potten, D. (1976). Etsha: A successful resettlement scheme. *Botswana Notes and Records* 8, 105–119.

Prins, B. (2006). Artisans and the tourism industry. PhD diss.

Prügl, E. (1999). *The Global Construction of Gender: Home-Based Work in the Political Economy of the 20th Century.* New York: Columbia University Press.

Santa Cruz Feminist of Color Collective. (2014). Building on "the edge of each other's battles": A feminist of color multidimensional lens. *Hypatia* 29(1), 23–40.

Sasibal, Kimis. (2016, March 7). From Kitsch to Craft to Economic Growth. *HuffPost.* Retrieved April 10, 2019, from https://www.huffpost.com/entry/from-kitsch-to-craft-to-e_b_9387292.

Schilar, H. (2020). "Arctic" tourism representations and the voices within. Via 18. https://journals.openedition.org/viatourism/6370.

Silva, R. Breda, Z., Brandão, F., & Costa, R. (2020). An analysis of the shopping behaviour of international tourists: The case of Porto. *Advances in Tourism, Technology and Smart Systems,* 10.1007/978-981-15-2024-2_55, 641–654.

Spivak, G. (1988). Can the subaltern speak? In C. Nelson and L. Grossberg (eds.), *Marxism and the Interpretation of Culture,* 271–313. London: Macmillan Education.

Stearns, P. (2001). *Consumerism in World History.* London: Routledge.

Swain, M.B. (1993). Women producers of ethnic arts. *Annals of Tourism Research* 20(1), 32–51.

Tanga, P.T., & Maliehe, L. (2017). An analysis of community participation in handicraft projects in Lesotho. *The Anthropologist* 13(3), 201–210.

Terry, M. Elizabeth. (2001). An overview of the Botswana handicraft sector. *Botswana Notes and Records* 33, 129–143.

Tlou, T. (1976). The peopling of the Okavango Delta c. 1750–1906. In *Proceedings of the Symposium on the Okavango Delta and its Future Utilisation. Botswana Society, National Museum, Gaborone, Botswana* (pp. 49–53).

Tlou, Thomas, & Campbell Alec C. (1984). *History of Botswana*. Gaborone: Macmillan Botswana.

Tucker, H., & Boonabaana, B. (2012) A critical analysis of tourism, gender and poverty reduction, *Journal of Sustainable Tourism* 20(3), 437–455.

UNWTO. (n.d.). Sustainable Tourism. Website. https://www.unwto.org/sustainable-development.

van Tonder, Louis L. (1966) *The Hambukushu of Okavangoland: An Anthropological Study of a South-Western Bantu People in Africa*. PhD diss., University of Port Elizabeth.

West, G., & Dube, M. (2000). *The Bible in Africa: Transactions, Trajectories, and Trends*. Leiden: Brill.

Wilkins, Hugh. (2011). "Souvenirs: What and why we buy." *Journal of Travel Research* 50(3). https://doi.org/10.1177/0047287510362782.
Wilkinson-Weber, C.M., & DeNicola, A. O. (2016). *Critical craft: Technology, globalization, and capitalism*. London: Bloomsbury Academic.
Wisudawati, N.N.S., & Maheswari, A.A.I.A. (2018). Potential of silver craft product through to community-based tourism sustainability in Celuk Village. *International Research Journal of Management, IT & Social Sciences* 5(1), 9–15.
World Tourism Organization. (2008). *Tourism and handicrafts*. UNWTO.

NOTES

1. Correspondence from Mary Jo Arnoldi, Curator, African Ethnology, Smithsonian Museum.

2. The weavers are ethnically Hambakushu originally from SE Angola. Their language is Thimbakushu.

3. Description on the Culturecrafts website.

4. Central Kalahari Game Reserve, Chobe National Park, Linyanti Reserve, Makgadikgadi Pans National Park, Mashatu & Tuli Game Reserves, Nxai Pan National Park, Okavango Delta & Moremi Game Reserve, and Savuti Channel.

5. I am using pseudonyms for individuals cited in this chapter.

6. Hambakushu is the ethnic group and peoples, Thimbukushu their language.

7. Government "drought relief" was a Labour Based Relief Progamme focused on "rural job creation" based on work on village projects for a daily wage (Jacques 1994).

8. Segola and Yoffe, no date.

9. The World Bank categorizes countries by their GNI per capita. The categories are low income: $1,025 or less; lower middle income, $1,026–$4,035; upper middle income, $4,036–$12,475; and high income, $12,476 or more.

10. They wove and sold their baskets to local shops, on the roadside, and at times, to the weavers.

11. According to its website, www.bedia.co.bw, BEDIA is a private sector organization charged by the Botswana Parliament with "identifying market outlets for locally manufactured products."

12. Correspondence from Mary Jo Arnoldi, Curator, African Ethnology, Smithsonian Museum.

SECTION III

Equitable Alternatives

Chapter 8

"My Mother's Recipe, My Nation's Narrative"

Intersections of Food, Militarism, and Masculinity in Maisa's Kitchen

Lindsey B. Pullum

Maisa has turned her sleepy residential street and modest home into the most popular eatery in the village. Off the main road that leads most visitors to the famous Druze Saturday Market, her restaurant is part food stop, part cultural museum. With a long, bricked parking spot for 40-passenger buses, the neighborhood transforms daily into a small tourist hub. The summer humidity is relentless in Daliyat al-Carmel as tourists quickly begin to sweat on their short trek from the bus to the backyard. They exhale in gratitude as their drops of sweat are met with a blast from the air conditioner inside. For a group of thirty, middle-aged Israelis, they move quickly to find their seats and ice-cold water. As they walk in, few notice the enlarged portrait of Sheikh Amin Tarif that hangs directly in front of the opening doors. No, this portrait of the Druze's foremost spiritual leader throughout most of the twentieth century goes unacknowledged.

The coffee stand however garners much attention. On the northern wall, fabric designed with the Israeli flag drapes down from a window ledge. The fabric is held up by a brass menorah and a metal tank, while a significantly smaller Druze flag is off to the side. Displayed with prominence next to the Israeli flag fabric is a certificate of kosher status, important for any Jew who might adhere to kosher food laws. These displays will soon fade from tourists' attention once food is served, but for the time being, their function is

Figure 8.1. Coffee Station Display.
Source: Photo by Lindsey B. Pullum.

unambiguous. The stand encapsulates the dominant narrative of brotherhood and patriotism told about a sect of Israel's Arab minority.

The relationship between the Druze people and Israel is often told with self-evident truth and definitive simplicity, but upon closer investigation can be categorized as a "minority of a minority" caught between colonial subjectivities that underpin Israeli society. Since Israel's statehood, funding for tourism development was granted to "good" minorities, like Druze and Bedouin communities, who demonstrated substantial allegiance to the Israeli state (Stein 2008, 78). Original efforts to draw tourists to Druze villages hinged on food markets and Druze cuisine, as certain Arab-Palestinian spaces for consuming food were deemed too dangerous (physically and ideologically). By the mid-1990s, in a time of peace following the Oslo Accords, Muslim and Christian Arab villages became a space synonymous with Jewish leisure and tourism. This shift left Druze to redefine and distinguish themselves from the now newly accessible spaces of Arab food tourism.

Druze women find themselves operating within this Arab-Israeli discourse of difference, in addition to navigating the internal norms about the role of women in Druze society. Touring Druze villages, specifically for food, opens up an avenue for engaging with and performing the Israeli nation by Druze women, whose generational knowledge of cooking Druze food make them especially equipped to create new culinary tourism experiences. In the largest

Druze village of Israel, Maisa has the only female-owned and operated restaurant; moreover, her restaurant is the only kosher one in a town flooded with hungry Jewish-Israelis. In this chapter, I provide an ethnographic study of culinary tourism situated within the colonial context of Israel. I see food and culinary presentation as part of national discourse, subject to a nation's ethos, values, and institutions. In Israel, military service, religion, and the Arab-Israeli conflict are the largest social forces that shape national discourse, and food responds to these forces. Druze women working in culinary tourism engage with the Jewish nation and conceive of themselves through a process I call *culinary militarism.* I focus on the ways Maisa adapts Druze food and her restaurant to align herself (and the larger Druze community) with the Jewish-Israeli nation. I first discuss her decision and her commitment to keeping kosher. I then cite the curated experience of the physical dining room as a space for performing belonging through paramilitary narrative and texts. Finally, I highlight how she articulates difference between Druze and Palestinian food through language. Maisa and her matrilinear network use feminine knowledge in conjunction with dominant foodways to connect to and then commodify military discourse for culinary tourists.

CULINARY TOURISM AND COLONIAL BELONGING

As I start my shift at Maisa's one afternoon, two Jewish-Israeli women timidly enter the restaurant. As was the case with these two women, regardless of the many warnings on Google to call ahead for a table, many people often stumble onto/into the restaurant with no reservation. Maisa's hospitality knows no boundaries, even when inconvenienced by unexpected arrivals. She promptly accepts the two women. As I set their place settings, I ask what brought them in. There are plenty of restaurants off the main road to choose from with similar offerings, better parking, and closer proximity to shopping. The women confidently reply, "We want the traditional Arab food." I expand on what I thought she meant by "traditional Arab food" with descriptions about freshness and local, seasonal farming. I use the colloquial Arabic names for the dishes, translating them into English afterward to add a touch of authenticity. *Bamiah* for okra and *malfoof* for stuffed cabbage leaves. Utterly unfamiliar with where anything is grown in relation to the restaurant, my job is to ensure these women know they are eating *Arab* food. These guests, not unlike many others, have "authentic, traditional" food in mind. The visitors who come to Maisa's admire its uniqueness; they do not want what they could make at home or find in another Arab village.[1]

Today, Israel's Arab minority makes up 20.95 percent of the population (not including the Palestinian-occupied territories of the West Bank and Gaza

Strip). It is composed of Muslims, Christians, Bedouins, and Druze, who only make up 9 percent of that number (about 2 percent of Israel's total population). Druze are religiously and politically different from, but ethnically, culturally, and linguistically like other Arab-Israelis. The state of Israel has reinforced Druze separate status from others in the Arab minority, designating them as "non-Arab Arabs," implying a separateness while still drawing on the discursive framework of the Arab-Jewish dichotomy. Although Druze do not feel entirely Arab or entirely equal in the Israeli state (Nisan 2010), their ability to create a meaningful Israeli identity explains how Druze of Israel remain both inside and outside of the Arab minority.

Israel's ethnodemocracy privileges the ethnic and religious identity of Jewish citizens over its Arab minorities. Such a "democracy" establishes a dichotomy of citizenship and belonging predicated on ethnicity and religion (e.g., Jewish or Arab). Despite extensive links to the greater Middle East and the Arabic language by Jewish-Israelis who immigrated to Israel after 1948 and Palestinian-Arabs who remained after/in spite of forced expulsion, Israeli identity privileges the Hebrew language and embraces traditions of Judaism and military service. Israeli-Druze citizens complicate Israel's ethnodemocracy as their intersection between the two national/cultural discourses of identity (Arab and Israeli) place them in a contradictory place.[2]

A fallout from Israel's colonial rule, "Druze particularism" was manufactured to separate and alienate religious sects within the Arab minority remaining in Israel after 1948 (Firro 1999). Bureaucratic distinctions (e.g., identity cards, educational systems) and military service rewarded Druze for being Druze (not Arab) and opened opportunities to the community that were inaccessible to Arab Muslims. As I have argued elsewhere (2020), this military service is a critical component to feelings and performances of belonging. But it is restricted to only men as decreed by the Druze community in the 1950s. Women have a prized position in Druze society; as cornerstone of Druze culture and life, and they must be protected from harm.

As the only group from Israel's Arab minority mandatorily conscripted to the army, Druze men prove to be the most integrated minority into dominant Jewish-Israeli life, with indistinguishable bilingualism between Arabic and Hebrew and strong relationships with Jewish Israelis who may otherwise have never met a Druze/Arab Israeli before. Post-military service, Druze men work alongside other Jewish-Israelis in security, travel, or entrepreneurship. In comparison, Palestinian and Arab-Israelis are denied access to certain jobs as they do not have the de facto military service of Jewish and Druze applicants. While military service is critical for Druze men to integrate into Israeli society, military and paramilitary discourse is a tool for Druze women to link themselves to the state in ways otherwise inaccessible.

I began dissertation fieldwork on tourism to Druze villages of Israel in 2015. I worked alongside a retired Druze Israel Defense Forces (IDF) Officer, whose second career was in Human Resources, recruiting and placing Druze at jobs in the area. Periodically, he would speak on behalf of the municipal government for groups interested in touring the village. Tagging along with him meant chaperoning groups of Army wives, tech-engineers, and second-generation Holocaust survivors as they walked through the Old City before stopping at a place for food. As far as I had seen, tourism to the village began with an exploration of who the Druze were (in general and to Israel) and ended with food in some way by the end of a visit. Whether you visited the Saturday market to shop for cheaper produce or paid respects to the local Yad Labanim museum for Druze soldiers, visitors were enticed to stay for lunch or grab *knafeh* (a cheese-based dessert) before descending the mountain.

While food tourism[3] encompasses the motivation of most tourists to Israel's Druze villages, I prefer Long's (2004) more expansive culinary tourism to describe the experience of my field site. Long defines culinary tourism as "the intentional, exploratory participation in the foodways of an other—participation including the consumption, preparation, and presentation of a food item, cuisine, meal system, or eating style considered to belong to a culinary system not one's own" (20–21). While Long highlights the role of the individual culinary tourist who voluntarily chooses to experience different foodways "for the sake of experiencing that change, not merely to satisfy hunger," making them an "active agent in constructing meanings" (21), there is certainly agency on behalf of those who are prepare, present, and contextualize food for culinary tourists. Culinary tourism invites scholars to consider more than ingredients, but also the labor, production, economics, and staging of food. I extend Long's culinary tourism to specifically question the linguistic, paratextual, and theatrical elements of food in context of a national "other."

Food studies scholars have long argued that the performativity of food is used to create and maintain national identity (Wilk 1999). Raviv (2015) connects food and performance within the Israeli context in order to understand how constructions of and connections to homeland are made. Following Kirshenblatt-Gimblett's performance studies approach, food and performance converge in three ways: doing, behaving, and showing (1999). "Doing" applies to the production of food products, dishes, and kitchens, and links economic interest to the physical land of the nation. "Behaving" applies to all customs, laws, and habits that define everyday food habits and how they establish socio-cultural practice and reflect political and ideological choices. Lastly, "showing" or displaying applies to creating national symbols via food in much the same way the flag or anthem might. The broadness of performance studies aids Raviv in her argument that food is a performance, just as a national identity is also a performance, and that the two work to create each other. I conceive of food and foodways as part of national discourse, using citations of the nation and being a citation of the nation. As with their socio-political identity, Druze food is caught between the markers of identity that makes it both Arab and Israeli.

Food in Israel is critiqued for its internal power imbalance and the efforts made to create a national cuisine from hummus[4], oranges, and barbeque.[5] To categorize food as distinctly Israeli, Lebanese, Jewish, or Arab is problematic given the historical multiculturalism of Israel's immigrants, in addition to the process of estrangement foods underwent in order to become "Israeli."[6] Within the Israeli context, national ethos and narratives of belonging are continuously conflated with similar understandings about ethnicity. From early in Israel's history, ethnicity was conceptualized through a push-pull dichotomy between Zionists and Arab regimes (Behar 2007). By contrast, Lockman's (1996) Dual Society model suggests that Zionism and Arab nationalism were "mutually formative interactions." Jewishness was conflated with Zionism and opposed to Arab identity and nationalism (although this was hardly the case for most Arab Jews). Israel's colonial ideology sought to rid Arabness from its citizens and establish a European-like culture, reconstituting Orientalist paradigms as a means of relating to Western Society (Khazzoom 2003; Shenhav 2006).

Historically, Israelis have been resistant to the influence of Arab culture on cuisine. Arab influence has often been minimized or attributed in other ways to Jewishness. For example, Mizrahi food (or food of Jews who immigrated to Israel from around the Middle East) is now "embraced by members of all the ethnic groups in Israel" (Avieli 2018, 23). The influx of Mizrahi Jews to Israel in the 1950s introduced Middle Eastern food practices to European Jews. Even though Oriental food practices were frowned upon, they eventually become mainstream through "a process" that Rebecca Stein describes "predicated on a disavowal of the Arab provenance of such tradition, whereby Mizrahi cuisine was marked merely as Jewish immigrant fare" (Stein 2008, 111). The Arabness of their food was re-branded instead to be another version of Jewish food (Avieli 2018).

For many of my Druze interviewees who cited food as a primary motivation for tourists to the village, they could not articulate a difference between their food and other (Palestinian) communities' versions of Arab food. In fact, when asked about the similarities between Druze and Palestinians/Arabs, many interviewees listed language, cultural traditions, and food as the same across both groups. With no significant differences between Druze food and Palestinian food, tourism to the village must draw on other markers, however ephemeral, to create difference. Culturally and linguistically similar to

Arab-Israelis, Druze are most different in their mandatory conscription to the Israeli Defense Forces.

I could not resist the opportunity to shadow Maisa and her sisters in the afternoons and evenings at the restaurant. I was quite familiar with the noise coming from her backyard as I spent one afternoon next door interviewing her nephews about their service in the IDF. During my interview with Abeer, I eventually stopped my recorder and demanded to know, "What is that noise?!" Abeer laughed before taking a long drag from his cigarette. He explained how Maisa started her own business a few years prior and hosted large groups who take up all the good street parking with their buses. To cut down on the noise of groups eating and talking (and clanking their plates), his family built a large fence with green shrubs to block the view and minimize the sound. While inconvenient to the surrounding families, Maisa's enterprise has no doubt financially benefited the extended family and engaged women from the entire street in working together.

A middle-aged woman, a white scarf wrapped gently around her brown hair, shuffles into the driveway from the suburban street outside Maisa's house. I have never seen her before today, but it is clear from her greeting with Maisa they are well-acquainted. Her light skin tells me she works indoors, not browned from working outside like her son who works construction. In her arms rests a plastic bag bundled at the handles. Fully packed, she walks into the kitchen to unload her plastic bag filled with folded pieces of bread. She has come bearing perhaps the food that most distinguishes Druze cuisine from other food prepared by Palestinian Arabs: *saj* (large, thin, flat bread). As opposed to smaller, fluffy pita sliced in half to make falafel sandwiches, Druze *saj* is most commonly served with *labneh* and *zaatar*, folded over itself multiple times. This type of bread is seemingly simple; it is thin enlarged dough baked on a large dome cook-top, from which it gets its name. It cooks relatively quickly but requires an artistry to stretch the dough to its paper-thin quality.

While quick to manufacture, Maisa's time as a food maestro for large groups cannot be spent on such a simple task. And so she outsources this labor to her neighbor. The neighbor's compensation is meager and does not compete with her husband's salary working security in Haifa, but it brings her pride knowing it is she who provides the bread hundreds of tourists eat at the most popular restaurant in Daliyat al-Carmel. Resourcing local and matrilineal/familial women's strengths is a staple feature of Maisa's enterprise. From nieces, sisters, daughters-in-law, and female neighbors, nearly her entire kitchen runs on the ingenuity and skill of women.

Empowering women is a fairly significant move by Maisa, whose generation has women working primarily as homemakers. Druze women have seen themselves in various occupations depending on age and state of the country.

In the early post-war years, Druze were still fairly poor with agriculture as the main source of income. Women gathered olives in nearby fields or gathered water at *Ein As-shoqef*, a local well. As Druze integrated more into Israeli society via the security services and business, the economic status of entire communities changed. Women of Maisa's generation became homemakers, caring for children and households. As economic conditions further improved and women's access to education increased, younger generations of Druze women moved into the public sector, taking jobs in offices and schools. Today, Druze women are more educated than their male counterparts who must serve in the IDF immediately after high school. Women are now pursuing careers previously unthinkable, like computer science, architecture, and broadcasting. Perhaps the most visible example of this generational trajectory is Gadeer Mreeh, who was a Druze Israeli broadcaster before serving in the Israeli Knesset as a member of parliament from 2019 to 2021. Maisa's own great-nieces only volunteer sparingly at the restaurant due to their studies at the local university. Their assistance is only solicited when food preparation or group size demands the presence of extra hands.

On an already-humid July morning when I walk over to help with food prep, the nieces and great nieces are huddled around a bag of grape leaves that has been dropped off. Daliyat al-Carmel is named for the grape leaves that surround the village, growing wildly between houses and along fence lines. With hundreds of open leaves that need to be trimmed and rinsed, the group of eight women sit around the open bag taking turns grabbing a handful of leaves at a time. Scattered scissors around the table point to the process. Take a leaf, snip off the stem, fold flat and stack. Each woman has a stack of her own. We are not competing to see who works fastest or whose stack is tallest; we are sitting quietly, enjoying the simplicity of the task and the company of family.

Arab Hands . . .

Getting Maisa to sit down for an interview is a big deal. This woman is running an incredibly busy restaurant, handmaking most of the food herself with varying amounts of help from her sisters. When we sit down at her kitchen table, I can tell taking things slow is not a mode she particularly enjoys. As a restaurateur who is constantly fielding calls for group reservations, it is not often I see her fast-paced, no-nonsense demeanor break. But that afternoon in her home kitchen with cups of hot tea in front of us, her voice gets warmer and her face softens as she describes her mother.

"Where did you learn to cook?" I ask Maisa. She smiles slightly but answers matter-of-factly, "My mother taught me to cook." There is a "*Duh*" vibe to her retort. I stare at her with a big grin, waiting for her to say more.

She senses my silent probe and elaborates (slightly), "I learned to work from my mother and my grandmother. We [my sisters and I] learned to help them in the kitchen." Maisa's contemporary recipes are actually a familial transmission of foods made from her childhood. One of the reasons she works so well with her sisters is that they each know the recipes and techniques from growing up together. They rarely speak during food prep, having memorized the mechanics and ingredients behind each dish they sculpt with their hands.

Maisa's feminine transmission of culinary knowledge passed down from her mother is framed as a secret but available to the world through contextual narratives grounding Druze cuisine. Subtitled, "A Taste of Secrets from the Druze Cuisine by the Cook from Daliat el Carmel." Maisa has a thin pamphlet for tourists with recipes written out. The recipe book has eleven recipes in total, from *warik aineb* and *majadra* to entrees and dessert. The contextual blurbs that foreground the recipes highlight matrilineal knowledge and technique. In her recipe for Saj Pita, she writes, "Us Druze mamas open the dough with our hands, while twisting it on the back of our hands, but you're probably better off using a rolling pin. The 'saj' is a deep frying pan that is heated and used on the outside curve (convex) to bake the pita." Hernandez and Sutton consider this "handmade" practice of cooking as a "modern strategy to valorize the tradition," but is actually the product of habit, fostered by familial transmission (Hernandez and Sutton 2003, 37). Maisa and her sisters share this memorized habit in how they prepare and cook food, with no formal recipes anywhere to reference. Tourists find this handmade tradition attractive.

Despite home/handmade food being a crucial part of the experience, Druze food is still markedly different from Israeli or Arab food. As I described to the couple in the restaurant, food is made fresh and from locally-sourced locations. The fight over land rights between the State and Druze draws attention to the Israeliness of okra and grape leaves. As the Carmel Mountain National Park expands to surround Druze villages, more land assumed to be Druze is being claimed by the state. However, because the ingredients are processed by Druze hands in the traditions of Druze/Arab/Middle Eastern women, the food becomes detached from traditional Israeli narratives that privilege land and working with *ha-eretz* (the land) as a sabra (a Jew born in Israel) or on a kibbutz (communal settlement for farming).

One of the foremost ways Druze villages have become popular to Jewish Israelis is through their food practices. We see this in other Arab villages like Abu Gosh and Nazareth (Stein 2008). In the minds of local Druze, food is cited as a key reason why people visit Druze villages so often. But even when Druze food and culture is marked as different from Israeli Jewish cuisine, it goes through a practice of becoming "kosher" to perform belonging.

I highlight the ways in which kosher Druze food is a construction that brings in Jewish mouths and garners their favor.

"So how did you decide to open your restaurant?" What once was a slight nostalgia for her mother turns to immense pride as Maisa recalls her time under the tutelage of a renowned Israeli chef in the Haifa area. Formerly a kitchen staff member for one of Israel's elite hotel chains, her recipes were used to spice up several courses offered at the hotel restaurant. Maisa's take on Mediterranean and Middle Eastern dishes caught the attention of then-head chef at the spa and resort, who incorporated her style into the menu. With inquiries from guests about where they could enjoy more food like Maisa's, it was the head chef himself who said "Oh, [she] offers Druze meals in her home." False at the time, Maisa rose to the challenge and, within days of the chef's declaration, printed business cards and hosted the first guests at her new home-restaurant: the chef and his friends.

While this early home arrangement worked for most visitors, some were deterred due to keeping kosher. The decision to make the restaurant a kosher one originally struck me as odd. One of the distinguishing features of the Druze village is just how overcrowded it is on Saturdays when Jewish-Israeli tourists come from nearby Zichron or Haifa to buy from local shops and treat themselves to some Middle Eastern/Druze food. But the business Maisa lacks on Saturdays is an easy trade-off for the business she ensures by keeping kosher at her Druze restaurant. When I asked her how she decided to become kosher, she said it was a logical and economically sound decision. She explained,

> I have a kosher restaurant . . . well, because I can. People from the very beginning, from the first time they came here to eat Druze food, they were asking for kosher food. Now since I'm kosher, the religious people, they can come to eat and sit and see our food. After [becoming kosher], I could work with bigger institutions, like the army and the hospital. Sure, before I was making a lot of food, but I wasn't kosher. Now that I am kosher, I serve a ton of people.

Keeping kosher affords Maisa a competitive angle; she can serve bigger groups without compromising the dietary restrictions of the religious or conservative Jewish population. Having built a separate kitchen and dining room from her main house to ensure kosher law, she can service groups like international Jewish high school students, IDF military groups, or US delegates.

Culinary Militarism

It is another cold dark night at Maisa's restaurant and my back vibrates in pain as I sit down next to Maisa's eldest sister. We are sitting on the leather couch beside the coffee and tea station that looks a mess, with sugar scattered between small paper cups used for Arabic coffee. Unplanned, we sigh simultaneously. Turning to each other, we burst into laughter. It is a feeling only two women can know after hustling for three hours. Tonight, we catered a banquet for IDF soldiers and the relentless appetite of the company left our shoulders tight and feet sore. Cooking, serving, and cleaning up plates of food for the sixty or more soldiers would wear out anyone. The leather couch was our place to breathe and put our feet up before getting back to dishes and mopping the floor. After our shared sigh-turned laughter, the eldest sister and I go back to silence; I sip my coffee silently while she reads missed messages on her phone, stopping occasionally to show me pictures of her grandchildren.

The explicit practice of feeding military mouths is only possible because of Maisa's commitment to making Druze food kosher. I did not serve soldiers often inside the restaurant since its size limited possible occupancy (especially in the rainy season when outdoor seating was not possible). Instead, Maisa's cuisine was often outsourced and delivered by van to locations around the Galilee. This particular night had been a special occasion with a presentation for the soldiers on a projector and awards presented to individuals, including to Maisa, who received a certificate of appreciation from the company for hosting. Applause buried the sound of me emptying metal bowls filled with olive pits, used napkins, and discarded chicken bones. Maisa's restaurant is easily converted from cultural museum to a military mainstage for reviewing and valorizing soldierhood. In fact, it is Maisa's honor to host and place herself in front of (and within) Israeli Defense. Maisa, her sisters, and even I, as a busser and waitress, create a feminine connection to military discourse by feeding its members and staging the space in which they consume kosher, Druze food.

To best articulate how subnational minority women assert themselves within existing colonial discourses, I present my concept *culinary militarism*. Part of Israel's broader *repertoires of violence*,[7] *culinary militarism* is a tributary of performing within colonial discourses that privilege forms of violence and militarization. It engages with embodied acts of cooking and the sensorial involvement of taste. For Druze and their culinary tourists, cooking, cuisine, and curation ultimately result in aligning oneself with the state through hegemonic, militaristic values associated with Jewish-Israeli life. *Culinary militarism* describes the process by which food and foodways are a tool of non-state/military actors to consent to state violence by way of military support.[8] In the case of subnational minorities and their foodways, it is through *culinary militarism* they signify a unique form of belonging otherwise inaccessible due to ethnonational discourse surrounding colonial subjectivity.

If food and tourism are performative of the nation, then Druze may perform it successfully through *culinary militarism*. For Druze women, *culinary militarism* is empirically represented in three ways. First, Druze women cater to IDF and paramilitary agencies in the restaurant or through military ceremonies. This direct "food to army" contact supports and, in some instances, praises their military practices. But to serve the army, all food must be made kosher to accommodate Jewish soldiers who observe kosher status. This in turn, accommodates Jewish tourists who may also keep kosher. Second, when served to culinary tourists, kosher, Druze food makes use of material culture and personal narrative to signal a patriotic relationship to the IDF. A centripetal force linking Israel's Druze minority to the Jewish majority is created through restaurant objects and Maisa's testimony to convey a Druze version of Israel's sociomilitaristic values. Finally, *culinary militarism* by Druze women use language to obscure the distance between themselves and other Arab-Israelis.

. . . Kosher Mouths

"Ah, the moped." I smirk walking up to Maisa's in the late morning. The rabbi (and his moped) assigned to monitor her ingredients and light the cast-iron stove is already there. While a genuinely pleasant man to speak with, I cannot help but feel the self-aggrandizing around his role as kosher food gatekeeper. As the ordained rabbi that services kosher restaurants around the mountain, he holds immense power in Maisa's day-to-day operations. She coordinates his visits whenever a family or group make a reservation so he can be there to light the stove before she starts cooking. Speaking to me about the importance of kosher food laws, he leans against his moped in the parking lot outside the kitchen, "Yes, if I come and her stove is lit without me or she has people over on Shabbat, I have to shut her down and she will lose her status immediately. But [she] is a smart lady. She knows to keep kosher." Subtly, he hints at the reason for her success but also the power he has to end it.

Kosher food and kosher dietary laws are a fixture in Israel's culinary sphere for both non-observant and observant Jews. Israelis strongly prefer kosher food and this preference shows in Jewish restaurateurs' decision to make a restaurant kosher (Ben-Porat and Feniger 2009; Liebman and Katz 1997). With Israelis setting the standard, kosher food is ubiquitous in public, religious, and military spheres. The widespread production of kosher food is a symptom of the non-separation of synagogue and state; it serves as a marker of Jewish religion, but also Jewish nationalism (Raviv 2015). Since a combination of secular and observant Jews serve in the IDF, kosher is the default setting for food served to soldiers. Kosher Druze food is symbiotically beneficial, creating an exclusive revenue stream for Maisa and curating a positive experience with Druze for Jewish tourists.

Kosher Druze food can be understood as a metaphor for the social experience of fitting into Israeli society. Regarding kosher rules, Mary Douglas comments, "It would seem that whenever a people are aware of encroachment and danger, dietary rules controlling what goes into the body would serve as a vivid analogy of the corpus of their cultural categories at risk" (1972, 79). Similar to New York Jews in the 1930s, food and its consumption are framed in terms of ethnic hierarchy and safety. Tuchman and Levine documented eating patterns of New York Jews and discuss how they ate Chinese food from a place of social superiority as Jews had a higher social standing than Chinese immigrants. This social superiority "made Jews feel safe and comfortable in Chinese restaurants" (Tuchman and Levine 1993, 386). In comparison, eating non-kosher Italian food was uncomfortable, as Italians held higher status in Europe and held anti-Semitic feelings. Chinese food was socioculturally distant and non-threatening to Jews. Kosher Druze food, placed within the sociopolitical contest between Israelis and Arab-Palestinians, does not challenge the Israeli national project, but moreover, Israeli Druze do not subscribe or contribute to Palestinian nationalism.[9] Creating kosher cuisine is a way to eliminate danger and protect Israelis' ethos from the existential threat of Arab nationalism, making Maisa's food culturally and ideologically palatable to Jewish visitors. Kosher food is interdiscursive with Jewish ideals and at odds with narratives of Palestinian food and nationalism.

Maisa's acceptance of kosher standards diagnoses hegemonic forms of being, and her food is subjected to the values of Jewish Israelis first, and to the tradition of her mother's recipes second. Her cuisine within an Arab context and for its original audience would not be subjected to kosher standards. The Druze religion does not require any form of kosher or *halal*[10] dietary processes per religious ordinances but does restrict the consumption of certain meat. With no formal recommendations on food preparation, making food kosher is an active choice that aligns Druze with Jews. This choice is rewarded and recognized online with multiple reviews from tourists who cite the kosher food as the main reason they made the journey.

PATRIOTISM ON STAGE

Any thoughts of kosher foodways are abandoned as soon as you sink your teeth into the *warik eyneb* (stuffed grape leaves). The sour tang from the grape leaves blends with the savory beef-flavored rice. You rip a piece of Druze *saj* and dip it in hummus. Time seems to have stood still and yet enough time has passed for Maisa to enter the dining room. It is showtime.

After guests have had enough time to eat at least one or two plates full, Maisa makes her way from the outdoor kitchen to the restaurant floor and presents the story of her life. She knows the explanation (told in Hebrew, translated to English) of her own life and experiences being a Druze in Israel is likely to be the sole account tourists will hear about the Druze community, what Druze believe, and the relationship they have with Israel. She starts from her beginning: she was born and raised on the mountain and is part of the Druze religion. Her family, like many others, has lived in Israel/Palestine for centuries. Her father worked in Haifa at a bank under British rule, but she brags he was one of the first Druze men to serve in the army before Druze conscription officially began in 1956. Her husband also served, but she is most proud of her two sons, one of whom is a police officer in Haifa and the other an officer in the army.

Drawing on these familial associations to the IDF, Maisa frames herself as part of a lineage of patriots (even though Druze women are forbidden from serving in the IDF). Maisa is able to use her business and skills as a chef to overcome her exclusion from the military, the very institution that forges the greatest connection between Druze and the state. She does so first through kosher catering fit for IDF soldiers and second through the restaurant's staged material culture. The connection between Maisa and the military is entextualized in the surrounding material culture (Israeli flags, photographs, and testimonials from army servicemen); these artifacts create interdiscursive conditions necessary to take part in Israeli narratives of belonging.

Like a dinner theater, either at the beginning or end of every meal she serves, Maisa explains her biography and her sons' work in the military and police. She is a fervent supporter of the IDF and enjoys humble bragging about her sons' service. Maisa's narrative leverages the IDF as a central, cultural touch stone. Narratives of loyal army service work to round out the Jewish-Israeli experience of consuming kosher Druze food. Through this meta-theatrical space, guests are not only educated on Druze ideologies of military service, but the restaurant and Maisa's cooking is where she can herself embody of these ideologies. In front of guests, Maisa emotes the narrative of faithful service to the nation that aligns with larger national repertoires for belonging.

In May 2018, a guest praised Maisa's kosher restaurant and her ability to host members of security forces and ministry officials. In their online review of the space, the guest positively embraced Maisa's narrative of Jewish-Druze co-existence, recalling enthusiastically how Maisa's family resisted the "Arab gangs" of Palestine in the 1930s and 1940s. Her militarized narrative helps tourists separate Druze from other Arab-Israelis (in addition to concealing historical and recent Druze reluctance to serve the military) and instead aligns them with the secular and religious Jewish Israelis. These narrative associations of Druze military service are incredibly salient to Jewish audiences, conjuring national sociomilitaristic values. Nationalism, patriotism, and militarism are values reiterated from food production, to story, to virtual retellings.

Maisa's patriotism is supported by the surrounding material culture of the restaurant, which drives home the embodied values within cooking and storytelling. By thinking of the restaurant as a stage, artifacts on the walls and tables become props and sets that contribute to monologues offered by Maisa, her sisters, and guest lecturers made up of tour guides and retired Druze mayors. The props of the restaurant hold associative properties of the military. Like making Druze food kosher, signaling the IDF in a positive way is important in associating Druze with Israel. The stage and its peripheral texts point to a curated design, bridging the values of the state with the identity of the Druze.

Lining the longest wall of the restaurant are numerous photographs, collages, and official certificates of hospitality. Military accreditation and montages of Israeli soldiers and government officials formalize praise for Maisa and her hospitality. Repeated images and language of gratitude on behalf of security forces reiterates the state-sanctioned approval of Maisa's service. One such certificate of appreciation from 2011 reads, "Recognition and appreciation for hosting an evening for *Mela* Division. Officers of Mela Division-A," reiterating Maisa's ability to cater to and host entire divisions. Simple in nature, this state-approval of (military food) service highlights a unique avenue for Druze women to engage with the IDF, an institution otherwise inaccessible. Her self-narrative, military appreciation posts, and discursive markers (i.e., flags and tanks) decreases the socioideological differences between being Israeli and being a member of the Arab minority. Using conventional and creative conventions associated with the military, Maisa encourages tourism audiences to connect the staged pieces with militaristic values. But her ethnoreligious difference is always omnipresent. Even with kosher food, narratives of allegiance, and curated material space, tourists' attention is always somewhat focused on the ethnic contrast and conflict between Jews and Arabs. And so, Druze-Israelis must minimize the sociopolitical distance between their loyalty and their identity as a non-Arab Arab.

MUKHTALIFEEN MINHOM (DIFFERENT FROM THEM)

With the onset of cold and rainy weather, Maisa's menu changes slightly to accommodate the in-season vegetables and weather-appropriate tea. An omni-present drizzle lingers as the sun sets on the cold January evening, and I make my walk to Maisa's kitchen. Despite only being 6:00 pm, the sky is already black with an unrelenting chill. I am most concerned with this weather as I know I will be running plates of food from the kitchen to the dining room, which are only separated by ten feet. But it is ten feet in cold rain. I must stay strong. Once I arrive, all concerns over weather are completely removed as I find out the touring organization for the evening. Maisa tells me we are serving a group from America (not at all uncommon mind you). But this group turns out to be a bipartisan, government delegation associated with an American Israeli friendship non-profit. The organization's goal is an enduring and cooperative friendship between the two countries; programs are created to increase the American understanding of contemporary Israel in all its facets (not just the Israeli-Palestinian conflict). It is incredibly intentional to have the US delegates hear about the perspective and experience of Israel's ethnic-religious minorities from Maisa.

"You can tell them about me and the Druze," Maisa permits. "Huh?" I respond incredulously. Now aware of the delegation's visit, I am *mitragshet*. This Hebrew word captures the essence of feeling anxious, excited, and exhilarated at once. I only met Maisa and joined her restaurant barely two days before and now we were hosting a delegation. I feel my body start to warm from nerves. Maisa is unphased. Tonight, my culinary skills are completely irrelevant; it is my cultural background she will leverage. As an English and Arabic speaker from America, I can translate short interactions between herself and league members that occur outside the stage of what will be her formal speech. More than that, I can field questions about Druze-Israeli life, which ultimately fulfills the objective of coming to Maisa's in the first place. There is no time to debate my role (or insecurities about it); the 40-passenger bus has arrived and the voices of its members head in our direction. It is showtime.

The bi-partisan group of assembly speakers and state legislators are eager tonight. By the looks of some faces, eagerness is prompted by the desire to hear Maisa's story and learn more about the diversity internal to Israel. Others are prompted by the steam rising from the silver bowls of hot food. They are understandably distracted from Maisa's introduction by the smell of spices and juicy chicken. Working in tandem with the group's tour guide and translator, Maisa tells the story of how she went from cooking with her mother, to hotels, and to her own restaurant. She recalls her father working in Haifa before Israel's independence. She concludes with her sons' military careers. As she speaks in Hebrew, the tour guide translates to English for the group. The story of peace and assimilation is an enthralling tale to the delegation as they nod along and whisper, "Wow." in awe of the cultural co-existence experienced by Druze Israelis.

The one man sitting closest to the translator but farthest from the food seems to be mentally somewhere else entirely. His hunger has been made worse by the incredible smells of food out-of-reach. He appears to have stopped listening entirely to Maisa. His eyes dance from container to container, perhaps dreaming about what is underneath. While Maisa's oral history is a brief five minutes, this man is clearly dying from the wait. Soon Maisa's speech ends and the tour guide begins explaining the contents of the different food (again translating from Maisa's Hebrew into English). At the end of the foods' descriptions, a cacophony of scraping chairs erupts as the group gets up to form a line at the buffet. The individuals tear into their meal. It is not long before I scurry from the kitchen to the dining room with refilled water bottles and carafes of lemonade.

Then the real work begins; as I refill drinks, the curiosity of the US delegates is piqued. I am very clearly not related to Maisa, as I physically stand a foot above her with lighter skin and blue eyes. "So what do you do here?" one delegate from Oklahoma asks me. I am explaining my research on Druze tourism and Druze life in Israel when Maisa's sister walks up to me and puts her hand around my waist. She looks at me and asks in Arabic, "Are they enjoying the food?" I translate, "Are y'all enjoying the food?" Certainties of pleasure are spoken and I assure the sister they are loving the food. She leaves my side to refill another container of rice. The introduction of Arabic into the league members' experience has added an unexpected layer of authenticity. This Hebrew-English, Arabic-English, English-Arabic linguistic dance highlights the trilingual space Druze occupy in Israel.

Language is a foremost feature of identity and Maisa uses it to obscure similarities and differences to Arab-Palestinian Israelis. In the above example, Arabic is used to reiterate uniqueness and heritage as Arab-Israelis who live peacefully in Israel. And yet, Maisa's work with Jewish tour guides mandates she speak in Hebrew to the point where it is easier for her to use Hebrew than Arabic to communicate. Moreover, the texts in her restaurant are predominantly in Hebrew, including the recipe pamphlet she has out for tourists to take home. This bi- and trilingual, multi-cultural/ethnic balance reverberates through each component of Maisa's restaurant. She carefully straddles Arab heritage of the Druze religion and relationship to Arab-Israelis with Hebraicized life in Israel, creating Kosher, pro-military food and performances. Never leaning on one side for too long so as to lose her unique positionality. Her cooking style, staged restaurant, and linguistic performances are each an important feature to separating Druze from Arab food, but the differences must not too overstated to undo the curated experience of touring an Arab-Israeli village. Maisa balances a fine line of being just Arab enough to appeal to local and international tourists while promoting the culinary differences of Druze food.

CONCLUSION

This chapter examined culinary tourism through the hands of a Druze woman; it unveiled the ways national discourses of militarism and ethnoreligious identity influenced the conditions under which subnational women can engage with the state. Maisa utilizes the values, tools, and tastes of Israelis Jews to curate her home restaurant into the most popular destinations for food in her Druze village. As an advocate of the army and leveraging her sons' service in the IDF, Maisa combines family food knowledge with kosher practices and staged performances to create a space that cites forms of Israeli belonging, connecting her own intersectional identity into Israeli society.

Maisa has completed *culinary militarism* in full. She feeds the mouths of IDF soldiers, replicating and reproducing kosher food practices. Using material culture from the IDF, she intertextually aligns her narrative with the state and the military in such a way that she signals militaristic values for culinary tourists. Finally, she plays with her bilingualism to draw attention to her Arab identity, if only ephemerally, before code-switching into Hebrew.

Online reviews from summer 2021 echo these sentiments. One tourist commented it was "[A] great experience . . . and unusual because it is also kosher and therefore rare." Others praised the "excellent traditional food." Perhaps tourism to Druze villages is still predicated on food tourism, but Maisa's enterprise shows us that the experience in the room is truly the distinguishing factor between other locations for Arab food. Culinary tourism and curated experiences (from food preparation, staging, and language) maintain the "good Arab" brand long held by Israeli-Druze.

ACKNOWLEDGMENTS

This chapter builds on ethnographic research conducted between 2015 and 2018), which was generously supported by the Robert A. and Sandra S. Borns Jewish Studies Program, the College Arts and Humanities Institute, and the Office of the Vice President for International Affairs at Indiana University Bloomington. I thank Robin Valenzuela, who read through several drafts of this chapter and provided helpful comments to improve the chapter's organization and logic. I also thank Georgetown University, the Center for Contemporary Arab Studies, and the American Druze Foundation for the resources to continue research and publish on this topic during my time as a post-doctoral fellow.

REFERENCES

Ariel, Ari. 2012. "The Hummus Wars." *Gastronomica* 12(1): 34–42.

Asad, Cher and Lena Mansour. "Criminal Foods." *Palestine: In-Between*. Podcast audio. April 16, 2021. http://palestine.mei.columbia.edu/events-spring-2021/criminalfoodspodcast.

Avieli, Nir. 2018. *Food and Power: A Culinary Ethnography of Israel*. Berkeley: University of California Press.

Bardenstein, Carol. 1998. "Threads of Memory and Discourses of Rootedness: Of Trees, Oranges and the Prickly-Pear Cactus in Israel/Palestine." *Edebiyat* 8(1): 1–36.

Behar, Moshe. 2007. "Palestine, Arabized Jews and the Elusive Consequences of Jewish and Arab National Formations." *Nationalism and Ethnic Politics* 13(4): 581–611.

Ben-Porat, Guy, and Yariv Feniger. 2009. "Live and Let Buy? Consumerism, Secularization, and Liberalism." *Comparative Politics* 41(3): 293–313.

Douglas, Mary. 1972. "Deciphering a Meal." *Daedalus* 101(1): 61–81.

Ellis, Ashleigh, Eerang Park, Sangkyun Kim, and Ian Yeoman. 2018. "What Is Food Tourism?" *Tourism Management (1982)* 68: 250–63.

"Following Adalah Intervention: Israel Reformulating Ban on Harvesting Wild Herbs Used In." Adalah, August 20, 2019. https://www.adalah.org/en/content/view/9794.

Khazzoom, Aziza. 2003. "The Great Chain of Orientalism: Jewish Identity, Stigma Management, and Ethnic Exclusion in Israel." *American Sociological Review* 68(4): 481–510.

Kirshenblatt-Gimblett, Barbara. 1999. "Playing to the Senses: Food as a Performance Medium." *Performance Research* 4(1): 1–30.

Kuntsman, Adi, and Rebecca L. Stein. 2015. *Digital Militarism: Israel's Occupation in the Social Media Age*. Stanford: Stanford University Press.

Lockman, Zachary. 1996. *Comrades and Enemies: Arab and Jewish Workers in Palestine, 1906–1948*. Berkeley: University of California Press.

Pullum, Lindsey. 2019. "A Resolute Display: Culture, Life and Intersectional Identity in Israeli Druze Photography." *Anthropology of the Middle East* 14(1): 87–107.

———. 2020. "Faithful/Traitor: Violence, Nationalism, and Performances of Druze Belonging." Indiana University. ProQuest Dissertations Publishing.

Raviv, Yael. 2003. "Falafel: A National Icon." *Gastronomica* 3(3): 20–25.

———. 2015. *Falafel Nation: Cuisine and the Making of National Identity in Israel*. Lincoln: University of Nebraska Press.

Shenhav, Yehouda A. 2006. *The Arab Jews: A Postcolonial Reading of Nationalism, Religion, and Ethnicity*. Stanford: Stanford University Press.

Sidali, Katia, Pascual Morocho, and Edgardo Garrido-Pérez. 2016. "Food Tourism in Indigenous Settings as a Strategy of Sustainable Development: The Case of Ilex Guayusa Loes. in the Ecuadorian Amazon." *Sustainability* 8(10): 967.

Stein, Rebecca L. 2008. *Itineraries in Conflict: Israelis, Palestinians, and the Political Lives of Tourism*. Durham: Duke University Press.

Sutton, David E. 2010. "Food and the Senses." *Annual Review of Anthropology* 3: 209–223.
Ting, Hiram, Kim-Shyan Fam, Jacky Cheah Jun Hwa, James E Richard, and Nan Xing. 2019. "Ethnic Food Consumption Intention at the Touring Destination: The National and Regional Perspectives Using Multi-Group Analysis." *Tourism Management (1982)* 71: 518–529.
Tuchman, Gaye, and Harry Levine. 1993. "New York Jews and Chinese Food: The Social Construction of an Ethnic Pattern." *Journal of Contemporary Ethnography* 22(3): 382–407.
Wilk, Richard R. (1999). "'Real Belizean Food': Building Local Identity in the Transnational Caribbean." *American Anthropologist* 101(2): 244–255.

NOTES

1. This distinction articulates the highly problematic disassociation guests make between Druze and other Palestinian-Arab Israeli restaurants and villages. This disassociation is proof of the 'divide and conquer' approach to Israel's Arab minority; separating ethnically Arab Israelis into discernable religious factions is a means for subverting collective resistance to Israeli policies that discriminate against Arabs.

2. As I have written elsewhere, this place between discourses allows for creative and intersectional expressions of community that highlight and appreciate both an Arab and Israeli identity (Pullum 2019).

3. Food tourism is a primary means of relating to a destination and a people/culture, in addition to accessing more localized, indigenous forms of knowledge (Ellis, Park, Kim, & Yeoman 2018; Sidali, Morocho, & Garrido-Pérez 2016; Ting, Fam, Jun Hwa, Richard, & Xing 2019). The multi-sensorial experience of food tourism creates a lasting impression on tourists thus imbuing food and its creators with a high amount of power to shape and reshape the takeaways tourists leave with.

4. Most recently, food has found itself in the legal sphere with the criminalization of *zaatar*, *akkoub*, and *miramiyya*, common plant ingredients in Palestinian and Druze food ("Following Adalah Intervention" 2019). In a recent podcast episode of *Palestine: In-Between, Criminal Foods* (Season 1, Episode 2), civil rights attorney Rabea Eghbariah discusses how Ministry of Environmental Protection deemed *zaatar* a protected plant, banning its picking, possession, and trade (Asad and Mansour 2021).

5. Food scholars have many foods to choose from when looking for examples: hummus, falafel, oranges, and barbeques. Bardenstein, "Threads of Memory and Discourses of Rootedness" discusses the transition of oranges to symbolize Israeli modernity. Raviv, "Falafel" and "Recipe for a Nation" reviews the assimilation of falafel to Jewish culture. Avieli (2018) outlines the transition of barbeques from *Mizrahi* Arabness to Israeli-Jewishness.

6. Raviv (2003, 2015) is the foremost anthropologist on this topic. She outlines completely the estrangement of falafel from Arabs and its transition to Israeli culture, with popular folk songs of the 1950s incorporating it into the lyrics. It was further

assimilated into Israeli culture by way of kosher food laws (that encompassed Israeli life, including the IDF).

7. *Repertoires of violence* describes "the collection of embodied action, including sensorial engagement, by which individuals and communities participate in violence of and for the state, both within military contexts and civilian environments" (Pullum 2020).

8. I draw inspiration from Kuntsman and Stein's *digital militarism* and their analysis of digital communication platforms as militarized tools by the state and civilians.

9. Going to the Galilee for Arab food implies traversing a separate sociopolitical landscape for Jewish tourists. Druze do not present any political or ideological obstacles and have been deemed "good Arabs" in their ability to conform to Israeli politics.

10. Halal describes the process of preparing meat as defined by the Quran. The Quran outlines the particular ways in which you slaughter animals.

Chapter 9

Entrepreneurial Domesticity

Women on the Forefront of Touristic Endeavors in Rural Costa Rica

Karen Stocker

In a rainforest ecotourism destination in Costa Rica, women artisans have been central to touristic innovation for decades. Women in two cooperatives that began at the start of a 1980s tourism boom in their region took on the role of cultural producers. Externalizing their domestic skills for a nascent tourist market, household sewing tasks led to embroidering images of rainforest flora and fauna onto products to sell. Cooking turned into the basis for a jam-making cooperative, and later, a restaurant frequented by locals and tourists alike. Simultaneously, families began to host North American students, offering room and board, thus instigating another form of income that centered women and placed an economic value on their expected domestic tasks. The story of the beginnings of the women's cooperatives offers a case study in deliberate cultural production from the inside. The effects of the women's efforts have extended well beyond their initial goals and have led to community leadership roles, dramatic changes in expected gender roles, a microlending organization that builds upon the connections and trust they cultivated early on, and a business model that prioritizes wellbeing over profits, while still serving to generate income. Although the founders of one cooperative intended to create a business that would allow their daughters to break the societal expectation of economic dependence on a male partner, they were so thorough in their artful revolution that the lesson their children of all genders took away from the endeavor was one born of seeing their mothers and grandmothers be a driving force for change.

Not only did these women-led initiatives constitute entrepreneurship and leadership, from their inception they had a reciprocal relationship with scholarly studies. The type of research fostered under such an arrangement was participatory action research and was intersectional and decolonizing from the start.

This chapter is based on original ethnographic research spanning more than two decades, beginning in 1992 with the founders of two women artisans' cooperatives, with in-depth follow up in 2009, and regular personal visits and consistent connection from afar in the intervening years. An undergraduate researcher under the supervision of sociologist Ilse Leitinger in 1992, I conducted participant observation with one of the cooperatives as well as on excursions where women from both cooperatives gathered and conducted semi-structured interviews with members of both cooperatives. I detail Leitinger's project in the following sections, and highlight the importance of participatory research in the sustainability of the cooperatives. I note here that between 1992 and the present, I have stayed in regular contact with the women, first through letters and occasional visits, then communication facilitated by social media and nearly annual visits. On five occasions between 2007 and 2018, I returned with my university students on study abroad trips. We listened to women from both cooperatives recount their entry into tourism and heard updates on their circumstances. Indicative of this ongoing, long-term engagement and in keeping with the tents of decolonizing methods (Kovach 2009, 48), those individuals identified in this chapter, or their descendants, have reviewed and approved a draft translated into Spanish for this purpose. The chapter stems from research that has always focused on rural women involved in tourism as innovators and agents of change in their communities, and thus focuses on women as producers not as a new angle of study but as a matter of course.

VALORIZING DOMESTIC EXPERTISE: A JAM COOPERATIVE AND STUDENT HOMESTAYS

In the 1980s, in the early wave of visitors to what would become a long-lasting ecotourist destination, there was at first only one small *pensión*, then increasingly more family-owned accommodations opened for foreigners. But large-scale, foreign-owned hotels never took hold (see Stocker 2013 for a thorough explanation). Host families offering homestay experiences were a logical extension of this small-scale, locally owned lodging. These homestays were centered around women's roles and expertise in an era when gendered divisions of labor were the norm. Perhaps paradoxically, it was those economic activities rooted firmly in gendered household activities, once monetized for tourist consumers, that led to a blurring of those gendered divisions of work for subsequent generations.

Shumaker (1995, 479) speaks to the rise in homestay programs for student travelers in Costa Rica in the early 1990s, specifically as tied to language programs. Although Shumaker does not overtly analyze the gendered nature of this work, the placement of her insights in the "Research Notes and Reports" section of a special issue dedicated to gender in tourism draws the connection implicitly.

Indeed, this is gendered work in that homestays monetize mothering and its traditionally component parts: cooking, housekeeping, and homemaking. Currently, more formalized homestay programs in the region label employees as "Mama *Ticas*," an amalgamation of the English language pronunciation of singular "mom" and the Spanish word for all things Costa Rican, in the plural form. The job title makes it quite clear both that mothering itself is a principal part of the job description and that the role is one of bridging cultural and linguistic realms. And yet, this labor is obscured as such, redrawn as kinship.

A motherly role is also what drove the homestay profession to begin with; seeking economic routes was very much related to mothers' domestic task of providing for family. Women's earnings, considered supplementary or extra, went toward school uniform costs and snacks children could buy to take to school or purchase during recess, as schooling went from being considered a luxury to a standard consideration in household budgets.[1] It was the women—and doña Aracelly[2] among them, who insisted on this change. Although doña Aracelly only went through half of a year of elementary school herself, she saw to it that her daughters attended school. All of her children completed a full elementary school education, which was the standard amount of formal schooling available to children in their rural community at the time, and her middle daughter was in the first graduating class of the local high school. Her youngest daughter also graduated high school. My 1992 fieldnotes of her experience are peppered with her recipes and instructions for how to make tortillas, how to prepare empanadas, and how to carry out other kitchen tasks—they exemplify the labor of a *mama tica*, though she was my host-grandmother.

Doña Aracelly did not have the opportunity to complete elementary school. Her parents took her out of school to help her mother with domestic chores, to pick coffee, and to contribute to family finances. She married at age 25 and moved with her husband far from her natal home in a time when return visits would be extremely difficult. She arrived in the mountain village, sight unseen, and lived in the house of her in-laws, a living arrangement she despised. As a young wife in her in-laws' home, the chore of gathering water passed from her mother-in-law to her, the lowest ranking, newest member of the family. But the errand of collecting water from a local stream got her out of the house and encounter others in a similar situation. Decades later, in her grandmother and great grandmother years, she helped start a business that would use this same model: leaving home briefly to a place marked simultaneously by work, respite, and the potential for solidarity.

Before she was a great grandmother, before she was *doña*, when she was a mere new wife, Aracelly would go down to the river to *jalar agua*—to haul water from the creek, for use in the home in the years before indoor plumbing. There is no doubt that it was a chore. The heavy pots, with water spilling over, surely seemed to weigh more with each step of the way home, on narrow, uneven paths in snake filled grass. This very water would underlie their other chores. It was used for bathing those children that survived birth.[3] Homemakers used the water they hauled for scrubbing clothes clean on a hand-carved wooden washboard or in a cement sink; for mopping the floor daily; for cooking first breakfast, then lunch, then coffee, then dinner, and for mopping again, in between, after men in muddied boots walked through for each meal.

For Aracelly, before she was a great grandmother, hauling water was a burden mixed with freedom. In this time when women were confined to homes and rumored about if they visited others—even their own mothers—too much, she went down to the river to carry out the obligation expected of her. There at the river, she could enjoy a time where she would not bear violence or insult. Before water came to the home and her daughters were grown enough to help her with the housework, she went down to the river for a paradoxical mix of work and rest.

In a 1992 interview, doña Aracelly recalled that in her younger years she overheard a man on a bus say that the then-new, 1949 constitution conceded to women the right to vote. He said that with suffrage, their citizenship card, and the tendency of the bold ones to wear pants, women were equal to men. But she knew better; she knew well that men still earned more. And she, who went down to the river daily, *jalando agua*, earned nothing.

Since her circumstances had not allowed her to get a formal education, she could sign her name, but she found reading laborious. So, in later years, she insisted her daughters get an education. Before she helped found the small jam cooperative not only so that grown women could learn new skills, but also so that their granddaughters could earn their own money, she went down to the river, collecting the day's water, and sometimes ran into another woman of the village carrying out this chore. Before she saw her granddaughters start their own business, become multilingual, get university degrees, and travel the world—before she saw them far surpass the possibilities she opened for them in the jam cooperative, before she saw them flourish in the opportunities she carved out for them, and for which she never took credit—she went down to the river, to do the one job of a housewife that justified her leaving home.

Before she saw her great grandson be raised by father and mother, equally, she went down to the river, leaving smaller children in the care of older ones, for a moment without crying, fighting, or jarring noise. Before she was a great grandmother, the women of her generation had to ask permission of husbands to do, purchase, or say anything. But on her deathbed, it was her daughter-in-law—who had loved her *suegra* and who doña Aracelly made sure to both welcome into the family and value—who gave the permission for her to go.

Before she was a great grandmother, she went down to the river to *jalar agua*. There, drawing out the task, she reveled in the greenery, the susurrus of the river, the turquoise of the mot mot bird's forehead, the fiery red, ebony black, and soft yellow of the coral snake, and the emerald tails of the quetzal. Before her home had light, she marveled at the electric blue of the morpho's wing. There, at the river, filling the pots, she rested awhile, and dreamed up a different world for those to come. Before she was a great grandmother, while she still had little more than the liberation that came with the burden of *jalando agua*, she concocted a world where innovation and tradition, and family responsibilities and freedom, all made sense together.

Before she was a great grandmother, she began to dream it into being, there at the river, *jalando agua*.

Doña Aracelly was one of several founders responsible for generating changes in family roles, economic activities, and local leadership. Together, doña Aracelly and her neighbors began a business making fruit preserves and pickled vegetables. They learned business administration, secured the requisite permits, navigated bureaucracy. My 1992 fieldnotes record their primary goal as learning, followed by generating income and providing work for others. Along the way, they built solidarity, and were able to use their workplace for consciousness-raising sessions, in addition to lessons on business administration skills led by a Costa Rican leader. Funded by two local organizations, a feminist Costa Rican facilitator trained in community organizing visited women's groups in several communities outlying a tourist hub, following a model of Community-Based Participatory Research (CBPR) (see Schensul et al. 2015, 190).

In keeping with the goal of CBPR projects to generate social change (Schensul et al. 2015, 190), in the group meetings held in the jam factory and led by the facilitator, women discussed their experiences, challenges, and goals. They addressed areas of concern to their community and families. They built bonds that have been long lasting.[4] The model not only adhered to the tenets of CBPR, but also of feminist Participatory Action Research (feminist PAR), which Schensul et al. (2015, 192) explain,

> is rooted in social psychology, phenomenology, empowerment theory, and social construction of knowledge. It focuses on the co-construction of knowledge based on lived experience. Most feminist PAR scholar activists work with relatively small groups of women to assist them, using reflexive research techniques, and autoethnography, including performance ethnography (performing their lives) to help them reflect together on barriers to self-efficacy, personal and group achievement, and external structural and political dynamics that restrict their voices.

The women of the jam cooperative acted out their favorite activities and their domestic tasks in a guided discussion of gender roles and the realities of women's lives, within a meeting that also taught business skills, so that they might adapt to a changing economic landscape.[5]

Doña Aracelly's *comadre*, a mother of several daughters–some of whom married doña Aracelly's sons, making them doubly family–was also among the women who carved a different path for their daughters than was available to her. It was true also for doña Margarita, whose husband was perhaps the most hesitant among the group. She married him when she was 15 and for 45 years prior to the start of the cooperative, she had followed the regional and generational norms of staying home. But doña Margarita wanted other options for her daughters. Her husband, too, made changes, however begrudgingly. As a solution to the dilemma of the women in his family being eager to participate in the cooperative, while also wanting to follow gendered norms, the women convinced him to donate a corner of his land to their inchoate operation. This way, technically, his wife and daughter never had to leave home. In this manner, the women pushed norms from the inside out, on their own terms.

When I brought up his acceptance of what, for him, were dramatic shifts in gender roles, 17 years after first researching the cooperative that his wife helped found, he acknowledged the changes along the way with a sort of disgruntled nod. His mouth pulled taut at one corner as he nodded slowly, looking straight ahead, in keeping with the rules of decorum for his generation. His *compadre*, a respected elder, had asked both of us to sit down for an interview. He requested I interview his neighbor about knowledge of medicinal plants and veterinary home remedies. If not for the request of his peer, who insisted that we meet on that day, my last day in the community, he would never have received me when his wife was not home. So, we met out on the front porch, in view of passersby, our chairs a modest distance away from one another, lined up exactly parallel to one another, preventing eye contact. He had made enough changes to his views of gender roles to permit his wife and daughter to work in the cooperative but appeared to remain resolute about gendered norms that need not be upset.

German-born, US-trained, and Costa Rican resident sociologist, professor, and mentor Ilse Leitinger (1997a, 225) cites a founder of another nearby cooperative, but true also of this one, in noting that the women's cooperative "has changed men's attitudes . . . somewhat.'" Doña Margarita's husband speaks to the "somewhat." In the same breath, Leitinger (1997a, 225) writes of changes necessary to allow for the cooperatives to take flight. "One of [the neighboring cooperative's] founders noted that 'women were not used to taking on responsibility. They had been taught to obey fathers, brothers, husbands; they were not ready to make decisions or demand room for their own personal development.'" A founding member of the jam cooperative confirmed that this was the case for her group. Doña Nuria said, in a 2009 interview, that in years prior, women had been raised in homes where fathers made the decisions. Decision-making was among the skills they learned by working in a women-only space in their newfound business, guided by a facilitator, in keeping with the practices of CBPR.

It was not only the work *within* cooperatives that was marked by PAR and CBPR. So, too was research about them. The jam cooperative, begun in 1988, was already four years underway when I first did research there under Leitinger's supervision. But I witnessed the transformation of gender roles, leadership, and women's increased participation in community meetings in real time. Leitinger did not work through an established homestay program, but rather arranged homestays herself, drawing on her ample long-term connections in the region, her reputation as an engaged scholar, and her intrepid character. Leitinger worked with a local community planning organization to connect student researchers to sites and issues that the organization deemed necessary to include. The organization's goal was to plan strategically for the year 2020, which at the time, seemed far in the future.[6] Under the direction of this organization, and partnering with its Costa Rican leader and professional organizer, student researchers were sent to rural communities that would be affected by increasing tourism.

The organization made good use of student researchers to interview as many people in rural communities as possible to ascertain what role, if any, people in each outlying community wanted to have in tourism, whether to be included in it or shielded from it. Student researchers were also tasked with identifying community needs in the interest of connecting communities to local institutes, initiatives, or organizations that might help meet these needs, some of which were symptoms of encroaching tourism and others of which might be met by an infrastructure expanded for tourists.

My role as a student researcher was to engage in Freirean, problem-oriented, collaborative research that would now be called participatory action research (PAR) (Schensul et al. 2015, 191), but at the time went unlabeled. In the present case, the social problem in question was rapidly increasing tourism and its potential resulting social and economic ills. Tourism would affect everyone in some fashion, through resulting rising costs of land and food as well as the presence of tourists, but not necessarily positively or evenly from tourist dollars. The capacity building aspects of PAR (Schensul et al. 2015, 191) happened in the women's group meetings and other aspects of this model were carried out by student researchers, interviewing as many community members as we could in order to take into account everyone's view, however marginal or central the individual was within local hierarchies. The dissemination of findings to all participants in the process is another key feature of PAR (Schensul et al. 2015, 190).

This approach required that I convene two meetings—one as a sort of focus group early in the project to ascertain which needs the community deemed most urgent, and one at the end, following numerous one-on-one interviews. This was designed as a collaborative session for debriefing, disseminating research findings, and introducing local organizations that could support the needs community members themselves had defined. At the first meeting, a respected elder and community leader let me know that gender roles were not changed yet, when he suggested I leave the meeting I was supposed to lead, so that I might prepare and serve coffee to him and other participants. At the second one, the transitional moment of gender roles in the community was on full display.

When doña Margarita was 64, she went to the town meeting, even though a common theme among women interviewed in 1992 was that meetings were for men—or at least they had been until quite recently. For most of the meeting, Doña Margarita stood silently by her husband in the back of the room. The men of the village listed the community needs one by one. Doña Margarita stood quietly until there was a collective lull, implying that they had nothing more to list. In her interview in 2009, she told me that she used to be "terrified to speak," but that her involvement in the cooperative helped her overcome this. I recall seeing this firsthand in 1992, when, at the meeting, I saw her visible trepidation and recognized the nerves in her voice as she spoke the single word *teléfono*. The men nodded in agreement, affirming her contribution of this item that none of them had mentioned, although the closest public phone was a strenuous one-hour trek away. Long after the meeting had progressed to address the issue of the havoc wreaked on the dirt road by the rainy season, doña Margarita still stood tall, arms crossed, and smiling proudly.

As noted by several women in interviews both in 1992 and 2009, increased confidence is an effect of working with the cooperatives. Other gains included income, which led to having a greater say in household finances. Guelke (2021, 59) offers a similar finding in a South American setting. Newfound self-assurance led to unforeseen outcomes, too. Independence was a benefit noted by various founders. Speaking to the effects of her long-term participation, one founder explained in 2009, "The biggest change was one I made, and [to which my husband] adapted." Doña Nuria began "to make my own decisions, nurturing some of my own aspirations." She added, confidently, "If I want to do something, I know I can" and that this leads to "satisfaction beyond the money." She credits the cooperative directly for that independence. Another founder, too, spoke of independence. In the words of Edith, "Now women have a voice and more rights. Before, men made all the decisions. The man's word went. Now women have more options."

The household was not the only area where women came to exercise greater decision-making power. With newfound business experience, women also ceased to see meetings and local committees, as well as the decision-making that happened therein, as the sole realm of men. As a result, women became community leaders.[7] Writing of Panama in the 1960s, Swain, too, noted that "women are moving from co-op leadership to community roles" (Swain 1989, 102), thus the two Costa Rican cooperatives are not isolated cases. A shift in confidence was key to this. Doña Isabel listed not only self-esteem, but also skills learned, education, travel, and solidarity as effects of her work with the cooperative over the years. Increased confidence stemming from starting their businesses, and as a root cause of their heightened community leadership, is true not only for women in the jam cooperative, but also in another, nearby cooperative. Leitinger (1997a, 225), noted, "many women have gained confidence through their business success with [the artisans' cooperative] or on their own." Indeed, women from both cooperatives, interviewed in 1992 and in 2009, agreed. Confidence, along with the ability to make decisions, was central to participants' memories of participation.

Paving the way for the jam cooperative was a women artisans' cooperative, nearby in the center of the nascent tourism destination. Some of its founders also lived in the community of jam makers. In a 1992 interview with a jam coop founder, she recounted that it was in part because of the prior cooperative that they were able to push the boundaries in their small village.

ENTERPRISING ARTISANS AND CREATIVE SOLUTIONS TO LONGSTANDING GENDERED NORMS

Just as the women of the jam cooperative had to ask permission of spouses, so, too, did four of the eight founding members of the neighboring artisans' cooperative. One woman, re-interviewed in 2009, revisited the early days with palpable nostalgia. As her grown daughter sat within earshot, doña Ofelia told me about asking permission of her husband to participate in an artisans' cooperative. The twenty-three-year-old daughter, previously engaged in other activities and not ostensibly listening, turned toward us and wrinkled her nose in an apparent mix of incredulity and disgust. The daughter's surprise at the scenario reflects how thorough, from just one generation to the next, was the transformation wrought by her mother and other women.

Like the jam cooperative, many founders and members of the artisans' cooperative also had to contend with husbands' varied views of this dramatic shift in gender roles. Leitinger (1997a, 225) acknowledged the variety of responses by the husbands of founders of the cooperative:

> Some men have been happy with the artisans' success. Many husbands welcomed the addition of their wives' cash income to their own. . . . Others, however, did not want women to take time out from home and child-care obligations in order to produce, learn, succeed, and maybe even travel or have fun with [the cooperative]. Undoubtedly, men feared that women would become too independent and that men would lose control.

To explain those men who were early supporters, Leitinger (1997a, 217) offers, "as Costa Rica's currency, the colón, lost value through inflation, many men welcomed their wives' economic contribution." She alludes to the planning organization that directed topics and destinations for student researchers when she notes that the cooperatives began in a moment of social and economic strategizing (Leitinger 1997a, 214).

The work of tourism scholar Martha Honey (2008, 190) reflects this shift, as she writes about the artisans' cooperative founded in 1982, noting that most of its founding members "had never before earned an income" but began a cooperative with the goals of "education, empowerment, and raising living standards in the community." Leitinger (1997a, 217), too, writes of the varied and simultaneous goals of the founders: "As for women's empowerment, the documents establishing [the artisans' cooperative] speak simply of 'economic and social improvement' of artisans' families, with leaders consistently emphasizing that the social goal is as important as the economic one." Its social goals included "of empowering women—both its members and, increasingly, other women in the community" (Leitinger 1997a, 210). She goes on to note, "Other aspects of [the cooperative]'s social mission include its many offerings of non-craft training, through courses on subjects important to rural women, such as health, nutrition, household management, and education" (Leitinger 1997a, 226).

A common theme of interviews in 1992 included the concern that this new entry into a formal economy must not disrupt women's home-related roles too much. Just like the women from the jam cooperative, they needed to push the boundaries from within, on their terms. The artisans' cooperative was meant to promote learning as well as income generation, while not curtailing the ability to maintain one's household. Women could build upon the expertise they had from the household realm, such as embroidery and sewing goods and clothing, for a tourist clientele. Doña Teresa summed up the goals of the artisans' cooperative as "helping women," first, and making money, as a secondary goal, without too much disruption to household obligations to allow participants still to care for family.

To this end, members of the cooperative could make ready use of the milk truck route[8] that collected cans of fresh milk from the surrounding communities and delivered it to the dairy in a central location. Although delivering products in person was also an option, participants in the cooperative could both send their finished products on the milk truck, for drop off at the store, and also depend on the same system to receive their art supplies to make more goods. This was one way that members who were pushing gendered boundaries by earning an income could participate actively, even if unable to leave the home when caring for small children, or not wishing to disrupt the expectation that they stay home—in keeping with presiding views of what it looked like to be a good mother and wife.

The cooperative organizers' goals of education, empowerment, and a form of income that accommodated family life, gendered expectations, and the logistics of rural living were evident throughout the early years of study on these cooperatives, as well as in more recent years. In an interview in 2009, Doña Teresa, a founding member and leader of the artisans' cooperative recalled the early days of the cooperative. She would advise members who were thinking about quitting, as a result of overwhelming problems at home, to stay involved, and merely produce less art for sale. She considered that every time a member would go to the shop to turn in products, that trip offered a moment to leave the home situation and perhaps share concerns with a friend. And for those members who sent their finished products to the shop via dairy truck route, they could count on regular visits from doña Teresa, whose role included visiting the outlying communities.

Doña Teresa recounted the few hours spent in each community as a chance for people to discuss their problems and lighten their load. She explained, "Those three to four hours had been like a mental cure. By telling the problem to the group, in the end, everyone gave a bit of encouragement, and advice. [Each member] no longer carried that heavy weight back [home] again." She focused on the relationships formed as the basis of the cooperative and the solidarity they generated, noting that in these rural visits, "I came to commit myself to [the other members] intensely." Doña Teresa also explained the flexible model of the cooperative that allowed each member to make as much or as little as they wished. There was no minimum production requirement for membership. This let members adjust their work hours in keeping with the changing rhythms of home life. In another 2009 interview, a member expressed her gratitude for that model. During a time of illness in her family, she considered quitting, but in addition to finding the extra income useful, she could not bear to separate herself from the friends she made and the environment that she credited with her own increased self-esteem. The cooperative was as much a support system as a place of work, if not more so. Leitinger (1997a, 225) agrees, noting that the cooperative "has profoundly changed the way women see themselves and the way others see them. In fact, some suggest [the cooperative]'s social achievements outweigh its economic ones."

One of the cooperative's founders reaffirmed this function as a deliberate part of the business model when she recounted the coop's history and current situation to my university students in 2018. In response to a student question about how much each affiliate must produce to remain a member of the coop, doña Teresa paused in the middle of her explanation of what happened if someone diminished their participation severely. Countering the expectation that such a situation might constitute cause for dismissal, doña Teresa responded that in such a case, they would go pay that affiliate a visit, to make sure they were alright and had support for whatever they might be experiencing.

In the same conversation, with my students, doña Teresa explained how Leitinger, in addition to having made a sizable monetary donation, oversaw student research that was part and parcel of the organization's foundation. In doña Teresa's brief explanation of the cooperative's history, she included student researchers supervised by Leitinger, whose research results informed the business. Likewise, Leitinger (1997a, 215, 223) cites student research to note areas of study that both drew from and influenced the cooperatives they researched. For example, with an eye toward finding culturally responsive solutions, two student researchers looked at "possible causes of low productivity" and included

> 1. The artisans' difficult working conditions, such as inadequate workspace and large family obligations; 2. Women's traditional perception of craftwork as recreation, not as work that merits an income; 3. Women's tendency to think neither in terms of a cash economy nor of their own labor as deserving financial reward.

Because art constituted a non-remunerated activity linked to standard domestic roles, it was devalued. Guelke (2021, 57), writing about a different geographic context, acknowledges this root of women's tourism-related work in wifely roles. Writing about Costa Rica, specifically, Barboza Nuñez (2020, 279–80) points out the gendered and exploitative ceiling on women's labor within tourism, either in the realm of pleasure or domestic service. Also writing about Costa Rica, Horton (2018, 234), considers women's roles in tourism as logical offshoots from their "household roles of cleaning, cooking, and serving others." I concur with Horton's assertion but add on an active and deliberate component. While the Mama Ticas do work with in a domestic realm, and the cooperatives draw from domestic expertise, I assert that their labor is marked by agency. I see this labor rooted in domestic expertise neither as a limitation, nor as an obvious precursor, but as a springboard. This was how the first generation of women working in the realm of tourism could get in.

For the generation that used their homemaking skills as an entry point to work in tourism, this was always entrepreneurship, it was always leadership, and it always had a reciprocal relationship with scholarly study. Moreover, it had repercussions for presiding attitudes about those skills. As Guelke (2021, 160) found, "as formerly unpaid domestic chores like cooking and cleaning become paid labor, their value increases." One of the founders identified this benefit, too. In 2009, doña Teresa said that people buying their products "made us value what we made," and prompted them to value themselves. "We have learned to value what we have." Perhaps where these skills have remained devalued is in academic consideration of them.

PARTICIPATORY ACTION RESEARCH AND DECOLONIZING RESEARCH

Just as women doing this work in the realm of tourism altered the whole model of life for multiple generations, something about the way of studying it did, too. Leitinger enacted and taught a model of research that now would be called decolonizing, but at the time was just what she considered to constitute responsible ethnography. It was a model of research rooted in mutuality (see Sanjek 2015). Such models require that a standard academic timeline give way to local ones (Rodríguez 2015, 59–60; see also Kovach 2009, 124 and Schensul et al. 2015, 194). Leitinger required that students stay longer than the semester-long program[9] to be sure to fulfill this obligation. And the model of research she taught was collaborative research rather than territorial. Nobody owned the research. Each student scholar contributed one part to a larger goal, keeping in mind that getting stories told responsibly was the primary goal, as opposed to any one student's own academic gain.[10] The understanding that the women interviewed were the experts and would be respected as teachers, regardless of their own history of formal education, was paramount.[11]

In the community where I studied, women in the eldest generation had between half a year and three years of elementary school experience. For subsequent generations, completion of six years of elementary school was standard. In the artisans' cooperative, one founding member had a university degree. The common thread was that each woman had completed the number of years of schooling available to her at the time and in the unique circumstances of each person, however constrained by gendered norms of the era, realities of rural life, and the wishes of their parents. One woman, when interviewed in 2009, was still working toward her high school equivalency, little by little. This was partly at Leitinger's urging. As doña Teresa explained in 2009, Leitinger always told the women that they could start to get a formal education at any age and informed them about night classes for adults available locally. "Because of Ilse, various women went [to night classes]. They learned to read and write, at least, and to add and subtract. Many women didn't finish the degree, but they learned." With the help of her children, doña Teresa was finishing her degree, while still leading the artisans' cooperative. She expects to complete it by the end of 2021.

The foundational epistemological instruction of recognizing women as experts had effects on the projects, certainly, but also in offering a model for teaching research methods, now in its third generation, at least. Leitinger referred to the methods of groups she studied (including the cooperatives) as "*investigación-acción*, or action-oriented research" (Leitinger 1997b, 335). Speaking of the women's groups with which she studied, Leitinger (1997b, 335) explained,

> As they continue to fight for equality, Costa Rican women can build on the relative openness of their society and its people's willingness to engage in dialogue. They can also count on a noticeable rise in institutionalization of processes of analysis, treatment, justice, job training, consciousness raising, and development of self-respect and efficacy.

She asserted that this would allow "women a chance to make their own unique contribution toward changing the traditionally patriarchal top-down management into a new and more appropriate way of bottom-up decision making" (Leitinger 1997b, 335–36).

Although Leitinger did not use the term decolonizing methodology, she taught her students interview strategies that were in keeping with its underlying praxis, assuring that communities of study owned the research results (see Kovach 2009, 145). Interview methods for projects meant to inform the cooperatives and communities' long-range planning rested on open-ended questioning that requests that interviewees describe their experiences, their feelings, and respect interviewees' views and interpretations of their own experiences (see Chilisa 2020, 250). Chilisa (2020, 262; see also Rodríguez 2015, 59–60) concludes that enacting decolonizing methodologies entails inviting research participants to "[critique] the literature written about them; [bring] into the discussion [I]ndigenous knowledge," and "[enter] into a dialogue with the researcher on the researcher's questions of interest." Returning research to the community and holding community meetings for debriefing and transparency further upheld the tenets of decolonizing methodologies (in keeping with Kovach 2009, 149). Leitinger (1997b, 336), in her applied research, urged women seeking change in Costa Rica to

> continue to push for a gendered analysis of the diverse experiences and problem-solving practices of women in the many ethnic groups, nationalities, and different religions in this diverse nation, including an evaluation of the impact of various foreign influences, the type of agencies represented, and any possible contradictions arising from these influences.

Using these methods, Leitinger and students could support the goals of the cooperatives studied.

Mirroring the intersectional lens Leitinger enacted and encouraged, feminist scholars in the 1990s were urging researchers to attend to matters of gender, race, and social class. Lugones (2020) insisted that scholars address the ways that race, ethnicity, and social class inhere upon gendered experiences. This body of literature influenced tourism studies, specifically (see Swain 1989; Swain 1995; Henderson 1994; Henderson and Hickerson 2007; and Henderson and Gibson 2013). Leitinger and her students were responding. Efforts that would now be labeled intersectional, decolonizing methods and philosophies were well underway. The failure of academia to normalize forms of inquiry that center gender, ethnicity, and social class, evident in the continued need for special issues or edited volumes to foreground gendered concerns in shifting economic landscapes outside of North America, reveals persistent misogyny, classism, and racism within academia.

That women's contributions have not been central to most studies of tourism reflects nothing about the driving forces by women, themselves. They not only weathered drastic changes resulting from tourism, but led them, deliberately and conscientiously. Their effort was not merely one of adapting to a changing economy, defending their place as tourism charged in. Theirs was one of being quick to respond to outside forces, to make sure they had a role in driving that change and using it to the advantage of their respective families, communities, and the visitors, themselves.

The actions of women at the start of a nascent tourism industry were not momentary adaptations. They sparked changes that became part of the fabric of their communities. They are cultural producers—not just of their wares (jam and art as well as cultural immersion experiences), but of cultural change, intentionally wrought. Their changes took hold and became normal parts of life for subsequent generations of community members, of scholars who both benefited from homestays and informed the cooperatives, and of tourists. It was women entering into businesses supported by tourism, who ushered these changes into being, supported them along the way, saw their rocky transition, and transcended the initial limits presented.

UPDATES AND LOGICAL EXTENSIONS OF THE COOPERATIVES

The cooperatives still stand. One remains small-scale while the other expands gradually. In the last 15 years, the artisans' cooperative has added a restaurant—still drawing from the initial model of valuing domestic skills and externalizing them for profit—and it is no longer strictly a women's cooperative. As of 2021, 82 women and four men are active members. The cooperatives face and surmount challenges, including navigating bureaucracy to obtain a barcode, dealing with trademark and intellectual property concerns, developing new product lines in keeping with tourist tastes, and adapting to changing realities. The artisans' cooperative still sells hand-sewn, embroidered potholders, but also iPad covers; new products include assorted repurposed goods, both in accordance with a continued focus on sustainability common to this ecotourism destination and also a logical extension of resourcefulness drawn from the domestic realm. Volunteers and international students continue to play a role in these new developments. Founders continue to teach, and the collaborative, applied model of research ensures that businesses are informed by research.

It is not only research models and options for women for which these founders created that lasting change, however. It is leadership, in general, that changed, and the increased options for children, regardless of gender, that women's efforts spurred. In a 2009 interview, doña Nuria noted that among the unforeseen effects of her efforts toward starting a business are that her now-grown children are good partners, collaborating at home. They are skilled and conscientious community leaders. "I see that they are dedicated, and that it doesn't scare them to start new things," doña Nuria summarized. The founders' children have had options. They travel, they run businesses, they study, they maintain families, and they are equipped to adjust as needed in accordance with the world around them. Just like the founders of two cooperatives, community members continue to adapt, for the benefit of themselves, their families, and their communities.

The jam cooperative's founders offered a model. Some of the founders still work in the jam coop. Community members continue to house foreign students. As a rational offshoot of homestay for younger travelers and in accordance with global change, community members manage Airbnb accommodations. One community member worked for years coordinating a study abroad program. In these ways, domestic expertise and its next generation extensions remain key to touristic endeavors. The cooperatives' founders' experience also offered a springboard for something new. The solidarity and profound trust that the founders of the jam cooperative built have also developed into a community bank and microlending organization (see Stocker 2013, 154).

CODA

The homes in the rural communities surrounding the tourist center have had running water for decades now. Norms no longer prevent women from sitting on the local water committees. The primary reason to venture to rivers these days is as owner-operators of tourism companies on canyoning projects or nature hikes. Various scholars credit tourism with locals' appreciation of conservation (see Honey 2010, 442; Yamashita 2003, 109; Bruner 2005, 88). I found something similar interviewing the cooperatives' founders and other tourism business owners (Stocker 2013, 169). In 2009, doña Teresa recalled her mother-in-law cutting and carefully arranging rounds of tree trunks to form a watertight roof. It is in later years, after working for the cooperative, that she came to see the arrangement as an art. She recognized the artisanry in tending a floor of packed dirt, adding and spreading water in a circular motion to make the floor smooth and firm as cement. Returning to my first interviews, I can see that doña Aracelly always had such an appreciation. She knew, early on, that domestic chores, the expertise that came from enacting them over time, and appreciation for nature were ready companions. It was this knowledge that she and her colleagues, *comadres*, and co-entrepreneurs enacted when they contributed early on to the tourism scene as creators of change.

REFERENCES

Barboza Nuñez, Esteban. 2020. *Las playas imaginadas: Turismo, imaginarios y discurso colonial en Guanacaste, Costa Rica*. San José, Costa Rica: Editorial Arlequín.

Bruner, Edward M. 2005. *Culture on Tour: Ethnographies of Travel*. Chicago: The University of Chicago Press.

Chilisa, Bagele. 2020. *Indigenous Research Methodologies*. Thousand Oaks, CA: SAGE.

Guelke, Karoline. 2021. *The Living Inca Town: Tourist Encounters in the Peruvian Andes*. Toronto: University of Toronto Press.

Henderson, Karla 1994. "Perspectives on Analyzing Gender, Women, and Leisure." *Journal of Leisure Research* 26: 119–137.
Henderson, Karla A. and Benjamin Hickerson. 2007. "Women and Leisure: Premises and Performances Uncovered in an Integrative Review." *Journal of Leisure Research* 39(4): 591–610.
Henderson, Karla and Heather J. Gibson. 2013. "An Integrative Review of Women, Gender, and Leisure: Increasing Complexities." *Journal of Leisure Research* 45(2): 115–135.
Honey, Martha. 2008. *Ecotourism and Sustainable Development: Who Owns Paradise?* Washington, DC: Island Press.
Horton, Lynn 2018. "Community-Based Ecotourism in Costa Rica." In *Tourists and Tourism: A Reader*, edited by Sharon Bohm Gmelch and Adam Kaul, 227–238. Long Grove, IL: Waveland Press, Inc.
Kovach, Margaret. 2009. *Indigenous Methodologies: Characteristics, Conversations, and Contexts.* Toronto: University of Toronto Press.
Leitinger, Ilse Abshagen. 1997a. "Long-Term Survival of a Costa Rican Women's Crafts Cooperative." In *The Costa Rican Women's Movement: A Reader,* edited by I.A. Leitinger, 210–233. Pittsburgh: University of Pittsburgh Press.
Leitinger, Ilse Abshagen. 1997b. "Conclusion for an Action-Oriented Research Agenda." In *The Costa Rican Women's Movement: A Reader*, edited by Ilse Abshagen Leitinger, 333–337. Pittsburgh: University of Pittsburgh Press.
Lugones, Maria. 2010. "Toward a Decolonial Feminism." *Hypatia* 25(4): 742–759.
Rodríguez, Sylvia. 2015. "The Field At Home." In *Mutuality: Anthropology's Changing Terms of Engagement*, edited by Roger Sanjek, 45–60. Philadelphia: University of Pennsylvania Press.
Sanjek, Roger. 2015. *Mutuality: Anthropology's Changing Terms of Engagement.* Philadelphia: University of Pennsylvania Press.
Schensul, Stephen L., Jean J. Schensul, Merrill Singer, Margaret Weeks, and Marie Brault. 2015. "Participatory Methods and Community-Based Collaborations." In *Handbook of Methods in Cultural Anthropology*, edited by H. Russell Bernard and Clarence C. Gravlee, 185–212. Lanham, MD: Rowman & Littlefield.
Shumaker, Nancy W. 1995. "Democratizing the Tourist Dollar in Costa Rica." *Annals of Tourism Research* 22(2): 479–489.
Stocker, Karen. 2005. "*I Won't Stay Indian, I'll Keep Studying": The Effects of Schooling on Ethnic Identity in a Rural Costa Rican High School.* Boulder: University Press of Colorado.
Stocker, Karen. 2013. *Tourism and Cultural Change in Costa Rica: Pitfalls and Possibilities*. Lanham, MD: Lexington Books.
Swain, Margaret. 1989. "Gender Roles in Indigenous Tourism: Kuna Mola, Kuna Yala, and Cultural Survival." In *Hosts and Guests: The Anthropology of Tourism*, edited by Valene L. Smith, 83–104. Philadelphia: University of Pennsylvania Press.
Swain, Margaret Byrne. 1995. "Gender in Tourism." *Annals of Tourism Research* 22(2): 247–266.
Yamashita, S. 2003. *Bali and Beyond: Explorations in the Anthropology of Tourism*, trans. J.S. Eades. New York: Berghann Books.

NOTES

1. While K-12 schooling is considered both free and compulsory in Costa Rica, the costs affiliated with purchasing the required uniform and supplies, plus availability of schooling, sets up a rural-urban divide. See Stocker 2005 for a thorough examination of this and related phenomena in the realm of schooling in Costa Rica. This divide was more marked in earlier years, though it still exists. In 1992 doña Aracelly's grandchildren attended a one-room schoolhouse located in the community.

2. In keeping with the American Anthropological Association's Statement on Ethics and its urging to remain in conversation with interviewees surrounding their preferences for confidentiality, I am following the wishes of each person interviewed or their descendants in the use of pseudonyms or real names, as each person saw fit.

3. As if given in evidence of poverty-induced traumas, early in my research elder women still listed the number of children they bore as one large number—whether number of pregnancies or number of births—then qualified by how many survived. Doña Aracelly had twelve pregnancies. Eight of her children survived to adulthood.

4. For a more detailed discussion of the cooperative, see Stocker 2013.

5. See Stocker 2013, 152–53 for a more thorough description.

6. The organization itself did not survive that long, but the connections it fostered using a deliberately inclusive framework in the early 1990s did, indeed, help community members mitigate the effects of pandemic-induced isolation, including economic need resulting from a shift from dependence on tourism to an abrupt lack of tourists, and fostering an exchange system to allow families to stay afloat.

7. For a more thorough discussion of women's recollections of their gains rooted in the cooperatives, and for input from more of the founders, see Stocker 2013.

8. The dairy route was also how community members could get to the clinic, run an errand in the central town, or use the one public telephone. Without the milk truck, an errand would have required an arduous hike ranging between half an hour or two hours, depending on the community in question. The dairy route also offered a young anthropologist a way to get to a central location to check for mail and use the telephone. My own use of the milk truck route offered me embodied learning of how much of a lifeblood the dairy route was for rural communities. This insight came in exchange for receiving the heavy metal canisters, slick with condensation, and still warm in cool morning air, hefted up by the truck driver, then rolling them into place so that they might all fit, to make efficient use of time and truck space. In the case of the outlying community where I stayed and worked with the jam cooperative, it allowed transportation to a central location that otherwise took a one-hour mountain trek for a spry 21-year-old in hiking boots over dirt and rock roads, sometimes in heavy rain. For local women in what would have been deemed appropriately feminine footwear, it would have taken considerably longer. This experiential learning let me understand how a place so geographically close to a tourist destination, for people with rental cars or ready transportation, could feel so distant in practical terms for people who resided there.

9. In a letter dated December 1991, she addressed the students who would join her unique study abroad program the following semester: "Some advisors, I for one, will

urge you to return a service to the community that supports you and your research, if you address questions of importance to that community. Please keep your return travel dates flexible, so that you have the time to return to report your findings to your community."

10. This was true in realms other than studying the cooperatives, too. The following year, I returned to do research on the first women voters of Costa Rica, a project to which Ilse introduced me. The project was too important to be dependent upon one person's schedule, so I did one part of it, then passed along my insights for someone else to do another portion of this longer, collective life history project.

11. Leitinger taught her students, most of whom were middle class white North American women from more urban realms, to see the people we interviewed as our teachers, not letting us for a moment believe ourselves to be more educated than the rural Central American Latinx women we interviewed simply because we were working toward university degrees at prestigious liberal arts colleges. Such nonsense would not have been tolerated. The people we interviewed were clearly the experts, though we were the ones who would earn course credits for their insights. This power imbalance did not go unspoken.

Concluding Thoughts

Re-Centering Women

Frances Julia Riemer

Through a timely and important approach, authors of the nine chapters in this edited text address long overdue issues in the study of tourism. Their chapter-length monographs, built upon original ethnographic research, re-center women by analyzing both recognizable tourist experiences (Stephens, Kipp, and Falconer) and host perspectives (Bolles, Becklake, Maksymowicz, Riemer) through feminist anti-colonial lenses. The result is an upending of our traditional understandings of tourism and surprising and at times, disorienting takeaways.

The familiarity of travel suddenly looks peculiar. Stephens reframes women's solo travel as a transgressive act that claims equal access to public space while discounting the implicit disrespect shown to local communities and practices. Kipp argues that by reinforcing the intersections of gender, race, class, sexuality, a volunteer tourism program actually re/creates a tourism bubble that reflects an ongoing legacy of colonialism. Falconer turns sex tourism on its head by focusing on the interplay of women's casual sex and the postcolonial privilege of backpacking travel.

Shifting focus to the local, Bolles juxtaposes job creation with low skilled tourist work, and women's skin teeth practices that help them maintain a sense of self. Becklake illustrates how intersecting power relations unfold in processes of exclusion, domination, exploitation, and in small acts of resistance in and outside a Guatemalan Spanish-language classroom. Maksymowicz narrates local resistance to global tourism practices that are imbricated into the imposed narrative of Costa Rica's *Pura Vida*. Riemer (this author) examines the colonial legacy of women, handicraft production, and tourism in her examination of Botswana's basket industry, and unpacks the neoliberal, neo/colonial development model on which the work of the women artisans has been based.

Across these ethnographic chapters, authors depict tourism as neo-colonial project that unfolds around race, class, and gender, is mapped by histories of Global North and South, and is driven by global flows of finances, ideas, people, and images. Women occupy intriguing roles in these tourism scenarios. As Cynthia Enloe (2014) reminded us, those from the Global North are now sufficiently emboldened to travel solo and engage in sexual activity, while those in the Global South are most often portrayed as pliable, resource poor, and in need of saving. But as Maksymowicz (chapter 6) illustrates, like Maria and Oskar, they evade and resist, as they, "continue to endure and modestly build worlds."

Finally, in their chapters on equitable and sustainable projects owned and managed by women, Pullum and Stocker offer alternative ways of conceptualizing tourism. Each introduces women who draw on their assets to assert control of tourism locally. Pullum offers culinary militarism as strategy employed by subnational women to engage with the state, national discourses of militarism, and ethnoreligious identity. Stocker traces a research effort that supported the growth of two women's cooperatives and led to projects, including a restaurant frequented by locals and tourists alike.

Authors attend to their own situatedness in the research sites and the scholarship. They are in these chapters, as participant and non-participant observers, as travelers and tourists, and as feminist scholars. Several of the chapters are based on research conducted over decades, and all are informed by authors' personal experiences, location, and positionality. While each chapter author sheds light on a different facet of tourism, they all trouble tourism by employing anticolonial theories to illuminate women's experiences. Framing tourism as neo-colonial project exposes privilege—its assertion, re/production, negotiation, and to some degree, resistance—that underlies these cases. Authors argue that particular women (i.e., those from the Global North, including many of us), have the privilege to travel, to study, to engage in casual intimacy, to demand familiar comforts, to silence local discontent, and to steer local industry. They also write about local women drawing on their own power, developing their own capacities, and negotiating the tourism industry on their own terms.

WHAT WE KNOW NOW

Many of us had hoped that COVID's forced pause on tourism might bring a new normal to travel. Harkening back to questions posed in Babb's foreword, we contemplate whether we will default to travel-as-usual or if we will seek new, more sustainable and equitable means of knowing the world. And we wonder if local hosts will roll out the welcome mat once again or push back against practices that reproduce familiar hierarchies.

In re-centering women and local meaning, practices, and context in their ethnographies of tourism, authors offer implications for us as travelers and for the tourism industry writ large. They illustrate the repercussions of women travelers' indulging in short-term intimacies and racialized, colonialist desires. They make vivid the high cost of providing care and safety to tourists who seek self-fulfilling experiences abroad and reveal ways local hosts push back against practices that reproduce old familiar hierarchies. They also describe equitable alternatives, centered in and based on women's lived experiences, and highlight the importance of deliberate cultural production from the inside.

Taken as a whole, the text suggests that tourism can be more than the pursuit of bucket list destinations and hedonist indulgences. Instead, we can draw on the work of transnational feminists to imagine in a different kind of travel, travel that fosters interconnectedness while acknowledging the links among patriarchies, colonialisms, racisms, and feminisms. Rather than global feminism, which, to quote Grewal and Kaplan (1994, 17), promotes "a universalized Western model of women's liberation that celebrates individuality and modernity," transnational feminists focus on gendered experiences within and across the Global North and South, examine their interdependencies, and develop relationships based on shared goals (Enns, Diaz, & Bryant-Davis, 2020).

What might transnational feminism mean for our tourism practices? In thinking through alternatives, I'm not referencing solely the principles of sustainable tourism, where, according to UN Environment Program and UN World Tourism Organization (Global Sustainable Tourism Council, n.d.), "tourism . . . takes full account of its current and future economic, social and environmental impacts, addressing the needs of visitors, the industry, the environment and host communities." I'm also drawing on the literature on socially-conscious tourism, where tourism is regenerative and locally-focused (e.g., eating and drinking, shopping, interacting). A list of other buzz words, including "transformative travel and tourism" (Ateljevic 2022), "conscious travel" (Pollack 2015), and "hopeful tourism" (Pritchard et al 2011), have been coined by geographers of tourism to reframe tourism as a means of making the world a better place. Websites like ethicaltraveler.org offer ethical destinations and journeys, based on the destination's records on environmental protection, social welfare, human rights, and animal welfare. Much of the advice is straightforward: spurn orphanage volunteerism and ocean cruises, prioritize animal welfare. Yet individual best intentions are rarely enough, and in practice, all too often re/produce neocolonial practices (see Stephens, Kipp, Riemer in this text).

The state can also play an essential role in organizing tourism. In an attempt to steward the country's tourism initiatives, post COVID 19 shut down, for example, the government of Bhutan announced a daily tourist surcharge beginning in Fall 2022. According to the country's Foreign Minister of Bhutan and Chairperson of the Tourism Council of Bhutan, "our goal is to create high-value experiences for visitors, and well-paying and professional jobs for our citizens" (Tourism Council of Bhutan, 2022). Of course, fewer tourists do not necessarily translate to a qualitatively different, more equitable kind of tourism.

Chapter authors contribute to these new tourism initiatives by focusing specifically on women's knowledge, as contextualized in their particular social positions. To quote Harding (1993, 56), "Starting off research from women's lives will generate less partial and distorted accounts not only of women's lives but also of men's lives and of the whole social order." By employing an intersectional understanding of tourism and centering women in analyses of tourism, we also directly address "asymmetries of power . . . and rigorously engage in complex constructions of agency" in tourism initiatives (Kaplan and Grewal 1994: 438–39). Our hope is the chapters in this text prompt ongoing conversations about tourism practices and policies that do not replicate cultural and economic hegemony.

This re-imagining will require white Western feminists and other elite women to see our own privileges as linked to other women's oppression, and to recognize the model of missionary work—of saving poor women—as neocolonial apparatus (Enloe 2014). But this imagining will also require a commitment to local, and in particular women-initiated and women-owned tourism ventures, by travelers themselves as well as by state and non-governmental funders. Stocker and Pullum (this text) offer examples of tourism initiatives that are women-centered, women-owned, equitable, and sustainable. As contrasted with Riemer's chapter (this volume) on handicrafts tourism as missionary venture to *save* refugee women in Botswana, Stocker and Pullum portray women carving out their own territory in the local tourism marketplace.

Anthropologists situate tourism as a "global movement of peoples engaged in mediated activities with human consequences, where culture is produced and reproduced, and where power is organized, negotiated, and re/produced" (Riemer 2020, 3). But gender is all too often overlooked in studies of tourism and other dynamics of globalization. Anticolonial and transnational feminists provide a unique template for addressing this omission. They "map these scattered hegemonies and link diverse local practices to formulate a transnational set of solidarities" and produce knowledge that is embodied, self-critical, and coherent (Grewal and Kaplan 1994, 19). As editor, my hope is that the chapters in this text further the mapping.

REFERENCES

Alexander, M.J., & Mohanty, C. (1997). *Feminist Genealogies, Colonial Legacies, Democratic Futures*. London: Routledge.

Ateljevic, I. (2020). Transforming the (tourism) world for good and (re)generating the potential "new normal." *Tourism Geographies*, https://doi.org/10.1080/14616688.2020.1759134.

Enloe, C. (2014). *Bananas, Beaches, and Bases: Making Feminist Sense of International Politics*. Berkely: University of California Press.

Enns, C. Z., Diaz, L.C., & Bryant-Davis, T. (2020). Transnational feminist theory: An introduction. *Women & Therapy,* 44(1–2), 11–26.

Global Sustainable Tourism Council. (n.d.). What is sustainable tourism. Retrieved July 21, 2022, from https://www.gstcouncil.org/what-is-sustainable-tourism/.

Grewal, T. I., & Kaplan, C. (1994). *Scattered Hegemonies: Postmodernity and Transnational Feminist Practices*. Minneapolis: University of Minnesota Press.

Harding, S. (1993). Rethinking standpoint epistemology: What is strong objectivity? In L. Alcoff and E. Potter, eds., *Feminist Epistemologies*. London: Routledge.

Pollock, A. (2015). Social entrepreneurship in tourism: The conscious travel approach. Tourism Innovation Partnership for Social Partnership (TIPSE), UK.

Pritchard, A., Morgan, N., & Ateljevic, I. (2011). Hopeful tourism: A new transformative perspective. *Annals of Tourism Research*, 38(3), 941–963.

Riemer, F. (2020). *Front and backstage of tourism performance: Imaginaries and bucket list venues.* London: Routledge.

Tourism Council of Bhutan. (2022). Press Release: Bhutan will reopen borders in September with renewed focus on sustainability. Retrieved July 23, 2022, from https://www.tourism.gov.bt/announcements/press-release.

Index

About the Editor

Frances Julia Riemer is professor of educational foundations and associate faculty/director of Northern Arizona University's (NAU) Women's and Gender Studies Program. An educational anthropologist who has conducted ethnographic research in the US, Africa, and Latin America, she focuses on gender, development, and sustainable communities, change and issues of equity and access, and cultural difference and the social organization of community, school, and workplace.

Dr. Riemer is the author of *Working at the Margins: Moving off Welfare in America,* co-editor of *Qualitative Research: An Introduction to Methods and Designs*, and editor of *Front and Back Stage of Tourism Performance: Imaginaries and Bucket List Venues*, as well as author of book chapters and articles in a range of journals and texts. She is a Fulbright Fellow affiliated with the University of Botswana, and her work has been supported by the National Academy of Education, the Spencer Foundation, the International Literacy Association, and the Arizona Board of Regents Innovation Fund.

Dr. Riemer's current research and writing projects investigate decolonial feminisms and tourism, safari and community-based natural-resource management in southern Africa, the intersectional gendering of niceness, first generation college students' study abroad, and intersectional understandings of climate action. Dr. Riemer is a diversity fellow at NAU; she chaired NAU's Commission on the Status of Women and co-chaired the City of Flagstaff's Commission on Diversity Awareness. She organizes for progressive policies locally, and regularly speaks on women's suffrage, the Equal Rights Amendment (ERA), and gender and climate change.

About the Contributors

Florence E. Babb is the Anthony Harrington Distinguished Professor of Anthropology at the University of North Carolina at Chapel Hill. She specializes in gender, sexuality, race, and class in changing contexts in Latin America. Her most recent book projects include a monograph on tourism in postconflict Latin American nations, and essays and commentaries on gender and race in Andean Peru, based on her decades of research in feminist anthropology.

Sarah Becklake is currently a post-doctoral research associate within the Institute for Sociology at Leibniz University Hannover. Her research focuses on the relationship between development, security, and tourism mobilities, especially within the Latin American context.

A. Lynn Bolles is professor emerita at the Harriet Tubman Department of Women, Gender, and Sexuality Studies and affiliate faculty in anthropology, African American studies, comparative literature, and American studies at the University of Maryland College Park. Her research focuses on the importance of economic analysis and on the impact of class on women in the English-speaking Caribbean.

Emily Falconer is lecturer in sociology at the University of Westminster, where she writes and teaches on gender, tourism, and travel narratives.

Amy Kipp is a PhD student in the Social Practice and Transformational Change Program at the University of Guelph. Drawing from feminist geography and critical development theory, her research focuses on the ways in which care is practiced and experienced by individuals and communities across various scales.

Kris Maksymowicz is a PhD candidate from the department of social anthropology at York University in Toronto, Canada. His research attends to the affective resonances of tourism encounters in Puerto Viejo de Talamanca, Costa Rica, as emergent forces in the unfolding of novel infrastructures of difference. Located at the intersections of feminist, decolonial, and non-representational anthropology, his work experiments with the ethnographic form, interrogating the limits of ethnographic temporalities and what counts as anthropological reality.

Lindsey B. Pullum is the American Druze Foundation Postdoctoral Fellow at Georgetown University's Center for Contemporary Arab Studies. Her dissertation, *Faithful/Traitor: Violence, Nationalism, and Performances of Druze Belonging*, was awarded the annual Ben Halpern Best Dissertation in Israel Studies from the Association of Israel studies. It explores the way Israel's ethnonationalist ideology influences the sensorial experiences of tourism in Arab-Druze towns across Northern Israel and the Occupied Golan Heights.

Siân Stephens is a senior lecturer at Middlesex University where she teaches business ethics and cross-cultural management.

Karen Stocker is a professor of anthropology at California State University, Fullerton. For over 25 years, she has carried out ethnographic fieldwork in Costa Rica. Topics of her research have included women's cooperatives; Chorotega Indigenous narrative practices; ethnic identity and schooling processes; the first women voters of Costa Rica; banana plantation labor; tourism and cultural change; and emerging social movements in the urban sphere. She is the author of four books stemming from long-term, ethnographic research in Costa Rican communities.

www.ingramcontent.com/pod-product-compliance
Ingram Content Group UK Ltd.
Pitfield, Milton Keynes, MK11 3LW, UK
UKHW041836031026
15587UKWH00002B/805

9 781666 901061